T0305827

Health Efficiency

Health Industrialization Set

coordinated by
Bruno Salgues

Health Efficiency

*How Can Engineering be a Player
in Health Organization?*

Edited by

Marianne Sarazin

First published 2018 in Great Britain and the United States by ISTE Press Ltd and Elsevier Ltd

ISTE Press Ltd
27–37 St George's Road
London SW19 4EU
UK

www.iste.co.uk

Elsevier Ltd
The Boulevard, Langford Lane
Kidlington, Oxford, OX5 1GB
UK

www.elsevier.com

Notices

For information on all our publications visit our website at http://store.elsevier.com/

British Library Cataloguing-in-Publication Data
A CIP record for this book is available from the British Library
Library of Congress Cataloging in Publication Data
A catalog record for this book is available from the Library of Congress
ISBN 978-1-78548-311-0

Printed and bound in the UK and US

Contents

**Chapter 8. Therapeutic Education for the Patient over
75 Years Old Living at Home**   139

Justine DIJON, Marianne SARAZIN, Vincent AUGUSTO, Thomas FRANCK and
Régis GONTHIER

Part 5. The Health Network.   159

Introduction to Part 5.   161

**Chapter 9. The Evolution of the Economic
Model of the Health Network in France:
Challenges and Prospects**   163

Aline LEMEUR

Chapter 10. Primary Care Electronic Health Data:

Richard BIRTWHISTLE

Acknowledgments

I would like to thank all the professionals who, through their contribution, made these interactions possible and allowed a large part of their work to be published.

Special thanks go to Vincent Augusto, professor and engineer at the *Centre Ingénierie et Santé* (Center for Biomedical and Healthcare Engineering) at the *Ecole des Mines de Saint-Etienne*, at the initiative of the *Journées sur l'Efficacité des Systèmes de Soins* (Days on the Efficiency of Healthcare Systems), who had a "musician's" soul and included me in his projects!

Preface

Efficiency is an important component of performance measurement. It is the optimization of the consumption of resources used to produce a result. From this viewpoint, the idea of devoting one day each year to the efficiency applied to the world of health was born. Sponsored by the *Ecole des Mines de Saint-Etienne* (Saint-Etienne School of Mines, an engineering graduate school), today, it brings together researchers, doctors, and engineers to engage in the same reflection on this problem. There are more and more collaborations between doctors and engineers. However, the application of industrial engineering and operational research in health systems is still poorly understood by the medical world, and it seemed important to offer this medical community a meeting place with the world of engineering.

A secret world, yet indeed real, healthcare has been the focus of attention since science used it to break through the holy grail of life, and government stock markets began to feel the wind of bankruptcy.

The 1946 preamble[1] to the Constitution of the World Health Organization (WHO) defines "good health" as "a state of complete physical, mental, and social well-being and not merely the absence of disease or infirmity". Health encompasses a multitude of parameters whose individual and plural characterizations can intimidate the most experienced scientific minds. Angina reassures the most Cartesian, cancer afflicts the most informed, and dementia confuses the most philosophical. In all cases, disease evolution can escape logic and disrupt complacency placed in outdated

---

1 Source: https://www.ncbi.nlm.nih.gov/pmc/articles/PMC4119253/.

science. Before the WHO developed its definition of health, the conception of health had considerably evolved, and although medicine has existed for several centuries, it has only reached its full maturity in the last two centuries. It was in the age of Enlightenment that it began to become a very real science.

Already relying at the time on other sciences (for example, mechanical science with the microscope, chemistry with mineralogical knowledge), health switched from being a religious conception, almost artistic in nature, to a rational and objective conception of the human body.

The advent of our modern era has been accompanied by the very rapid development of medical science, and with it, emerging techniques have made it possible to cure more and more diseases. We have moved from an almost veterinary and fatalistic medicine to an elaborate and hopeful medicine. Based on a series of statistics, even cancer patients can hope to see their median survival rate get considerably increased and a future for themselves. The psychiatric patient has now taken off their straitjacket and can return to an almost social life. The cardiac patient can engage in sexual activity without the fear of cardiac arrest. Not content to merely relieve symptoms, doctors go much further: they anticipate illness! A simple healer who became a visionary, increasing the screenings available within the framework of preventive medicine, doctors have begun to hunt down healthy people with the goal of making them immortal.

If the doctor is now sailing in calmer waters, surrounded by libraries of knowledge and recommendations, they still have many questions and continue their quest for knowledge. The doctor incorporates more and more technology and beliefs into their practice in the face of an ambivalent body, which sometimes escapes any reasoning!

However, all this has a price, and the ideal collapses before this down-to-earth reality. The increase in technological developments has been accompanied by an increase in the cost of patient care. The robotization of surgery, the precision of medical imaging techniques, data computerization and the increased specialization of knowledge are essential elements of modern healthcare. It represents millions of dollars in investments for all health centers, and modernization cannot be limited to one! This technology has also led to the development of new professions such as radiography technologists, biomedical technicians and health data managers, which are now essential players in the support team of doctors and nurses.

French health centers, like most of those in countries around the world, are therefore increasingly struggling to find an economic balance. The question of financing health technologies is becoming increasingly significant, even if they contribute to the development of new knowledge. "Illness" is expensive. This reality is giving rise to numerous reports and an ever-increasing grip of politics on the caregiver. In our era of modernity where humankind is aiming for Mars, we begin to dream of a "herbalist" medicine that defies ethics to try to reduce its shortcomings. But there is a strong public feeling, and political powers face an impossible choice: limit access to better healthcare or risk life? A real "Rubik's cube" of the general conscience, this question agitates both the systems that finance and the systems that care; for all are patients in the making and current payers.

Also, the words "efficiency" and "optimization" take on their full meaning when faced with this question. From a topic of worldly conversation, these terms have become a health research issue. It is a question of preserving an avant-garde medicine while minimizing the cost of its management. The *Ecole des Mines de Saint-Etienne* has understood this debate and put all its knowledge from the industrial world at the service of health professionals. Exchanges are born from this approach that we invite you to discover in this book.

Marianne SARAZIN
September 2018

Part 1

Health Technology Coordination Approach

Introduction to Part 1

"Technological health": these are two terms whose nature might seem antagonistic.

Certainly, technology has revolutionized the curative approach in medicine: chemotherapy, imaging and surgical robotics have made it possible to identify the enemy and eradicate it. But does that make you healthy?

The definition of health advocated by the WHO introduces the word well-being into its conception. Nothing is more abstract than "well-being". From a spa session (hydrotherapy, massages, etc.) to listening to a composition by Mozart, from climbing Mont Blanc at minus 25°C to a few gymnastic movements on a mat, a cuddle with a dog, the purr of a cat, the warmth of a stable, the murmur of the wind in a forest, and even the resolution of computer programs: everything can be a source of "well-being" and provide the sensation of eternity.

The body is forgotten in well-being. A cancer patient may feel well, as may a manic-depressive psychiatric patient, or a disabled person. Therapeutics increasingly includes relaxation and fitness sessions in their general management. Oncology programs combine the help of beauticians, sports educators and clowns. Disabled people may benefit from equine therapists to make them forget that at one point, the body suffered and was in agony. It is a question of reigniting the desire for pleasure for a consciousness that the disease has strongly challenged, and these sparks take an increasingly important place in healthcare. They are now proven to be effective. This state of affairs perplexes people of science who are constantly finding receptors, hormones and neurotransmitters to explain this poorly quantifiable success. And yet it works! The average survival rates of patients managed by these programs are increasing, and the complications associated with illnesses are diminishing – proof that there is indeed something imperceptible that pulls the body towards its "spirit human" condition. A long way from technology!

And yet, how should we organize all this in the current context of a moving society clinging to its mobile phone, to the Internet, solicited from all sides, scattered over an increasingly vast territory? Relationships change along with their habits, and patients too, even if the substance remains the same and the WHO definition retains all its relevance.

Coordination between professionals thus becomes a central concern for all professionals in the healthcare system. The collective position is at the service of the individual where any action must be adjusted in coherence with others. The implementation of health pathway coordination often requires rethinking one's profession. The capacity to adapt, the relative renunciation of its place as the center of a process in favor of an element centering on a process, cooperation, subsidiarity and neutrality are all qualities that must be developed for everything to work. Also, interaction among professionals and, intersecting care programs, becomes a real challenge where the objective is to limit interference and reach the end of the process without harmful consequences for the patient. If the patient posts a remission quicker than expected, or on the contrary, an unforeseen complication arises, this could cause the whole organization to collapse in a succession of psychological dramas where the hospital no longer wants to hear from the local doctors, the nurse no longer talks to the occupational therapist, and the health insurance ends up punishing everyone under the pretext of social justice! An unwelcome prankster, the patient feels left to their own devices, and the fruit of this placement rots with the hope that despite everything, health will return.

"Social networks", enabled by new digital technologies, have shown how powerful it is to organize a world in a short time frame. This is thought-provoking, and the social networks can be used for the health coordination of a cure for this need for coordination. Whether they upset our conceptions to the point of rejecting them or on the contrary stimulate our creativity, they challenge us on how to coordinate a group and on the interference of technology in the narrower world of health. Something to meditate on!

1

Forward Vision of Technologies and Health Knowledge

As technologies penetrate our entire society, they profoundly change our worldview and our individual and collective behavior. This is particularly true in the field of health in the broad sense, including but not limited to medicine. While our awareness of these technologies' contributions to our lives and knowledge is acute, their financing is beginning to pose a problem because our economy is in crisis.

QUESTION.– What new performances of the health system do these technologies potentially bring, at the cost of what concessions, and for what new advantages?

This is the question addressed in the work of the *Conseil Général de l'Economie* (French General Council for the Economy), nourished by the experience – with a strong international component – of a group of researchers from a range of disciplines (medicine, technology, human and social sciences, and education), as well as practitioners and representatives of various institutions.

This chapter also offers some additional reflections, specifically on the new data that these technologies make possible to collect: conditions and difficulties in obtaining it, and analyzing and sharing health data.

Chapter written by Robert PICARD.

1.1. Issue: justifying the value of health technologies

Technologies are becoming increasingly important in providing healthcare. The impact on health professions is considerable: collective, multidisciplinary, remote work. Meanwhile, our view of the disease is changing due to major technological advances in recent times: access to genomes, molecular biology, etc.

At the same time, the lack of compulsory insurance is not being improved while the needs of an older population, where the chronically ill are increasingly numerous, are growing. A desire to control expenditure is reflected in a quest to control practices, through increased information on the functioning of the health system and practices, and through the codification of medical and surgical knowledge (recommendations of good practices, accompanied by a tendency to "embed" these recommendations in informatic tools and robots as *Système d'information et d'aide à la décision* [Decision Support Information System], robots). The growing legalization of a citizen's relationship with the health system, the result of excessive trust in the power of medicine (commitment to results) and a new distrust of institutions, tends to make these recommendations opposable. Are we not on the way to an implicit Taylorian industrialization of care provision, with the authorities playing the role of the design office for healthcare provision processes embedded in machines?

This context led to the launch of a mission entrusted to the *Conseil général de l'économie* (French General Economic Council) on the link between technology and knowledge in the health sector. This chapter presents the main lessons learned. The reflections it discusses were shared in a multidisciplinary working group. This group includes the *Haute autorité de santé* (French high authority for health, HAS), health institutions, health professionals, and technological research – *Centre national de la recherche scientifique* (CNRS, the French national center for scientific research), and medical research – *Institut national de la santé et de la recherche médicale* (INSERM, the national institute of health and medical research). They met from February 2013 to November 2013, in 9 meetings involving 30 participants, all contributors (presenters).

1.2. Which technologies for which health knowledge?

A technological race for in-depth medical investigation is currently being observed to obtain less risky surgical interventions. This is part of a positive history of the relationship between medicine and technology: technology has enabled more precise and reliable observations and more effective surgical procedures. More recently, through the elucidation of complex biological phenomena, it is possible to curb pathological progressions and to cure illnesses.

Today, the search for more in-depth and solid knowledge on the functioning of the human body and its pathologies seems to go beyond the sole issue of the safety and effectiveness of the decision and action. Another logic emerges: that of rationalizing and controlling the provision of care. The practitioner who must give a precise diagnosis, the industrialist and the manager who wish to earn money through working in the health sector: all of them dream undoubtedly of a more "scientific" medicine whose laws would be universal and proven. This would make it possible to decide "for sure" on the right treatment or intervention, while minimizing the risks associated with human expertise. In this hypothesis, technologies should serve to explain this new "scientific management" of healthcare provision and by loading it into decision-making or operating machines. There is however a risk of no longer being apprehensive of the human being and no longer hearing its complaints. Moreover, age and disability, which affect people globally, have so far not been well-served.

1.3. Contribution of technologies to health knowledge

The contributions of health knowledge technologies are extremely numerous and varied. In this chapter, we propose a classification in three fields which takes into account the interaction between pure technology and human knowledge with its sensitivities.

1.3.1. *Facts and actions*

DEFINITION.– This is the area where technologies help to collect data and accompany the act, or even carry it out in an automated way.

DEVELOPMENT.– **"Facts and actions"** is the field of sensors and surgical robots. Sensors are diversifying and are invading our daily lives. However, their use in health is not immediate: there is the reliability of the collection, the relevance of the data collected; the value of these data, the conditions of their validity, depends on the treatments to which they will be subjected; however, these are not necessarily known in advance. The intervention robots, for their part, find a less hegemonic place than that imagined in the recent past: tomorrow, they should accompany and secure the practitioner's actions close to the patient, rather than operating alone, or via a remote control. An extension of the surgeon's hand, robots will allow interventions in inaccessible and/or microscopic areas.

1.3.2. Representations

DEFINITION.– In this field, information technologies enable the organization and processing of the data collected, their coding, and the development of standard models.

DEVELOPMENT.– **"Representations"** is an area in chaos. Under the influx of masses of data ("Big Data") now available, of a tangle of descriptive knowledge and interactions, the traditional models seem to have become too rough, but nothing is available to replace them yet. The medical image, intuitive and rich, is more and more mobilized. However, it is diverse, voluminous and complex to structure. Its exploitation should be accompanied in the future by a concise expression of the useful knowledge it contains. This would make it possible to free oneself from the constraints of volumes and those related to the modes of acquisition. At the same time, the international community seems to be running towards a gigantic model of the human body, of the biological and physiological mechanisms, of which it is the seat, the overall structure of which is difficult to perceive. The coded elements that compose it, which today are segmented, are, and will be even more so, so numerous (notably with genomics) that even their statistical exploitation becomes problematic ("Big Data"). This statistic, which provides the best "evidence" of the effectiveness of medical solutions, only legitimizes some of the recommendations made to clinicians today. This raises the question of how they can be programmed into tools and used as legally enforceable injunctions.

1.3.3. *Emotions and intersubjectivity*

DEFINITION.– This third field corresponds to that in which technology is in strong interaction with the human and in the service of individual and collective practice.

DEVELOPMENT.– **"Emotion and intersubjectivity"** remains relatively poorly invested in the realm of health technology. Collaborative forums and tools are, however, at the source of new knowledge from current, professional or secular experience. The human and social sciences allow us to subtly approach human interactions with tools, especially in surgery, but also with immersion devices (virtual worlds, games) or simulations, developed for practitioners of all professions, but also for the patient, without putting them at risk. Companion robots are programmed with components that perform certain functions which aim to emotionally interact with humans by learning and imitating the reactions of their "masters". However, ethnography and ergonomics have so far hardly been mobilized to develop technical systems that really facilitate the mobilization of knowledge teams of humans in close interaction.

1.4. Learning transformations

Learning is transformed through new technologies and new pedagogies. Indeed, learning strategies are diversifying under the impact of digital technology, in the sense of taking into account both individual needs and new opportunities for collective interaction. This is particularly interesting in healthcare practices, which used to be separate and segmented, as they now tend to be collaborative with increased practice, by necessity, in delegating tasks and decisions. Training in surgical interventions can now combine face-to-face interaction, with the use of simulation techniques and remote participation, by mobilizing virtual games and virtual worlds. In medicine, as in other health training courses, massive online training courses are making their appearance. The obstacles to the dissemination of these new methods lie both on the part of teachers, whose practices they upset, and of trained persons, who can only integrate the proposed pedagogy if they have prior cultural and/or practical knowledge, which should be able to be evaluated.

1.5. Economic issues

The need to put technologies at the service of medicine, but more generally of care and prevention, is a matter of macro-economic and public health considerations. In this sense, it is necessary to constantly ensure their value and to base this estimate on solid analyses, which are still very often lacking. This value is also recognized by any potential payer, including the caregiver, patient, their insurer, etc. The economic evaluation of health technologies must also develop an understanding of manufacturing costs, which has hardly been the case so far. Then, it is necessary to clarify the nature and the stakes of the industrialization process of the sector. Many support, logistics and management processes, even those directly affecting healthcare provision, data collection and analysis resources, can benefit from an industrial approach. On the other hand, the industrialization[1] of the act of care resulting from the careless transposition of methods from other sectors, based on the automated dissemination of unproven knowledge, could well lead to health disasters: for many personal and singular services – the art of care – success depends, on the contrary, on the exploitation of recent results from the human and social sciences and on technologies adjusted to their needs by the practitioners themselves. These technologies are likely to have an interesting future, which urgently needs to be addressed. Another social and solidarity-based economy, based on these aspects, is emerging.

1.6. The data question

The value of data is in everyone's mind and particularly agitates health professionals. It is not a question of developing the whole of the problem, but of bringing a certain number of open questions in reference to the field "representation" that was previously proposed.

1.6.1. *Models versus data mining*

First of all, even though important work has been done for years on the representation of humans, their anatomy, physiology and pathology, by carefully linking the various descriptive elements (ontologies), there is little mention of how these links will be treated when overabundant additional data will lead to the emergence of new correlations.

---

1 This is described in the book by Bruno Salgues published in the same set, entitled Health Industrialization.

QUESTION.– Will ontologies be abandoned?

QUESTION.– Models are hypotheses of the functioning of the human body: How will they be transformed?

1.6.2. *Big Data*

CONCEPT.– Big Data, or mega data, means a set of data so large that it exceeds human capacity or even that of conventional computer tools in order to analyze and interpret them intuitively or even rationally.

QUESTION.– Can we improve public health with Big Data?

A second aspect, developed in a 2013 Organization for Economic Cooperation and Development (OECD) report, "OECD science, technology and industry scoreboard", highlights the limits of statistical approaches in a universe where the descriptive parameters of the subjects observed are of an order of magnitude more numerous than the size of the population observed. The craze for "Big Data" is based on the hope of using high quality data in a timely and comprehensive manner, and that it will be possible to improve public health, be it prevention, an improved health status or a better quality of life.

Opportunities to learn and create value from Big Data will depend on the statistically valid use of the information. The volume of data collected and their heterogeneity constitute a major challenge. Unfortunately, so far, most statistical approaches have been used at a time when sample sizes were relatively small, and acquisition technologies and computing power were relatively limited.

There are theoretical and practical limitations of statistics associated with Big Data in health for the evaluation of cause and effect relationships.

More specifically, problems arise when we want to identify cause-and-effect relationships using multidimensional observational data. This multiplicity of dimensions in itself poses problems of estimation, classification and visualization. The similarity of the processed data and their observability weakens the validity and confidence level of cause–effect relationships. All in all, limited to the field of statistics alone (paradoxically, with Big Data, the number of experimental units is much smaller than the number of parameters to manage), questions emerge about how to measure the strength of the evidence of a causal relationship and how to combine evidence to learn something from that relationship.

The multiplicity of dimensions of observational data poses estimation, classification and visualization problems. The appropriation by the user of Big Data, variables and models require new tools and interfaces. The challenge is that the results obtained can meet clinical competence, a condition for creating value in terms of care and public health.

1.6.3. Open public data (Open Data)

Etalab[2] is a public structure that coordinates the actions of government administrations and provides them with support to facilitate the dissemination and re-use of their public information. It contributes to their design and coordinates their interdepartmental implementation. Etalab has identified the main databases or main public data sets existing in the field of health and has made this panorama available for a dual purpose: to make this information available to all, and to publish, in open data, the file of this public data mapping, so that everyone can use it.

Article 47 of the French Open Data Act 2015 on the "*Modernisation de notre système de santé*" (modernizing our health system)[3], is entitled "*créer les conditions d'un accès ouvert aux données de santé*" (creating the conditions for open access to health data). Its objective is to create a national system of medico-administrative data, to create a National Institute of Health Data and to reform the procedures concerning the registration number in the national identification register of individuals. The challenge is to centralize data from existing health and medico-social databases, such as the National Health Data System (NHDS) and making it available on the principle of open data.

---

2 Etalab/SGMAP Prime Minister's Service, within the General Secretariat for the Modernisation of Public Action, in charge of opening public data and open government.
3 https://www.legifrance.gouv.fr/affichLoiPreparation.do?idDocument=JORFDOLE00002958947 7&type=contenu&id=2&typeLoi=proj&legislature=14.

A public consultation was launched in April 2014 so that everyone could comment on the completeness of this mapping and express their views on the usefulness, legitimacy and risks associated with opening these data. This consultation did not generate the expected enthusiasm. The "Blue Button" initiative launched in the United States to give American citizens access to their health data has not been as successful as expected.

QUESTION.– Open data for the citizen?

Honestly, this remains doubtful. So far, there is no question of this opening having any value. However, the conditions for this value to be shared and the modalities of this sharing remain an open question.

1.6.4. *Ecological data and Living Labs*[4]

Mobile applications and connected devices allow new knowledge of reality to be developed. The methodological benefits of the "Ecological Momentary Assessment" (EMA)[5] found in Table 1.1 summarize the characteristics of the evaluation, published in 1994 by Stone & Shiffman, and were rediscovered some twenty years later by Genevieve F. Dunton.

Characteristic	Comments
Ecological	Environmental elements can be taken into account, and the information gathered can be placed in the experiential context of "real life". This results in a detailed validity of the data collected or "ecological validity".
Temporary	The evaluation is based on a snapshot of the data. The approach favors real time, the "way to go" measurement, which avoids recall bias.
Evaluation	This approach has the following complementary characteristics: data collection is carried out by the individual person; measurement is repetitive, intensive, longitudinal; it allows analyses of physiological, psychological and behavioral processes over time.

Table 1.1. *Data characteristics*

---

4 More information on this subject is available in two books by the same author, Robert Picard: (Picard 17a, Picard 17b).
5 For more on this subject, see: https://www.ncbi.nlm.nih.gov/pubmed/18509902.

Today, the technologies to implement this approach are in common use: mobile apps allow such investigations in real time. Smartphones themselves are equipped with an integrated accelerometer, GPS, camera and video technology. The whole device can be synchronized with other ambulatory sensors via Bluetooth (for example, heart rate, respirators, air pollution, UV radiation).

CONCEPT.– The Living Lab is a research laboratory, bringing together public and private actors, companies, associations and individual actors, with the aim of testing "life-size" services, tools and new uses. The aim is to bring research out of the laboratories and into everyday life, often with a strategic view of the potential uses of these technologies by promoting open innovation, sharing networks and involving users from the design outset.

Living Labs allow both the design of such solutions in the service of an in-depth knowledge of reality and the evaluation of these solutions before and during the initial phase of their distribution. Many scientific fields must be included. It is a question of anticipating the contexts likely to constitute the environment for solutions, imagining them with future users so that they can project themselves there, putting them in a situation close to reality in simulated environments, followed by placement into a real situation. At each of these phases, the sociologist, anthropologist, designer and ergonomist proceed with the observation, analysis and formulation of recommendations, requirements and specifications. At the same time, living people will generate observation data and traces as soon as technical prototypes are available, with the opportunity to identify early correlations with interpretative hypotheses.

1.6.5. *Evaluation data: conditions for feedback*

The absence of comprehensive evaluation models is a reality regardless of the meaning of the different value propositions of health products: for the patient, the practitioner, the organization and public health. The amount and disparate nature of evaluation data are problematic and difficult to integrate. Experimental results are not comparable, which is an obstacle to public decision-making. The European model for the assessment of the telemedicine method (*MAST*)[6] of the Telemedicine Assessment Report provides an overview of the impasse that scholarly organizations and

6 For more on this subject, see: http://www.renewinghealth.eu/en/assessment-method.

international bodies may find themselves in without a research effort on the gathering of useful new information. The Commission, convinced of the need for a European collection of decision indicators, has so far limited its evaluation criteria to universally available measurement results, abandoning the idea of defining new, more relevant criteria, but whose measurement instruments have yet to be invented. Thus, the question of defining and collecting the new data needed for decisions to meet the new health challenges remains a difficult and poorly resolved question because it is a long-term one.

1.6.6. *Data qualification*

A final issue concerns the validity of the results of the data analyses. When using data, it must be ensured that the conditions under which they have been specified and collected are compatible with the processing they will undergo, especially if they are numerous and emanate from different sources and contexts. This relationship between a more or less documented knowledge of the context of data collection and the comparison of these data with potentially decontextualized hypotheses, not known at the time of collection, deserves to be studied. In addition to this question, in the field of health, the ethical and legal question of patient consent for such use of data is unknown beforehand.

1.7. Conclusion

Health system actors, to varying degrees engaged in the new digital world, have or could have at their disposal new and useful information to improve its overall efficiency. However, as the suggested prospective representation indicates, this information must be able to flow between the different actors, so that a global understanding of the health system as a system becomes accessible. An effort of structuring and articulation between the different actors of the ecosystem at work remains to be made.

For this to be possible, it is necessary that these actors share a vision of everyone's place in the global system and of their contributory value. The proposed model is one avenue for such a work. It also provides a key to understanding the nature of data and desirable exchanges within the system to create value.

In this perspective, involving users, both professional and laypersons, is a necessity. They are the main contributors to the value created. The information they generate must make sense to them in order to be valid. The diversity of situations and decisions to be taken in complex and singular situations must counterbalance an understandable desire for summary data useful for overall management. It is about sticking to the real, not the virtual, so that digital tools serve the clinic and the health of populations.

1.8. References

Picard, R. (2017a). *Co-design in Living Labs for Healthcare and independent Living*. ISTE Ltd., London & John Wiley & Sons, New York.

Picard, R. (2017b). *La co-conception en Living Lab Santé autonomies 2*. ISTE Editions, London.

2

Coordination Between Professionals:
a Public Health Issue

Coordination between health professionals is a key concept in today's healthcare and has clearly been lacking to the point that the French government has positioned coordination at the center of its *Stratégie nationale de santé* (national health strategy) from 2011 (*Ministère des affaires sociales et de la santé* 2013). Coordination is, however, an organizational mode that is *a priori* logical, generically defined by the "need to ensure the coherence of different tasks with a view to efficiency" (Bloch and Henaut 2014). Coordination must thus proceed from a collective position in which all actions must respond and adjust to two fundamental, yet contradictory needs: the division of labor into diverse and complementary tasks on the one hand, and coherence in the joint or parallel execution of these tasks on the other (Mintzberg 1982). One of the challenges in healthcare to ensure coordination between the various professionals, is the multiplicity of trades, expertise and skills needed to structure, prior to implementation, patient management. In France, a major problem is the coordination of interventions between the various care and service providers, which fall into three relatively compartmentalized sectors of activity:

1) *the health sector*, which includes institutional professionals, health establishments (mainly hospitals) and ambulatory care with primary care on the front line;

2) *the medico-social sector*, which includes accommodation establishments for people with disabilities or loss of autonomy, providers of

Chapter written by Gérard MICK, Mario DEBELLIS, Marc WEISSMANN, Michel SABY and Alexandra GENTHON.

home assistance services and institutions which regulate these services and manage financial aid, primarily with the services of the *départements*[1];

3) *the social sector*, which includes workers in social and professional integration and solidarity, including the *Centres Communaux d'Action Sociale, CCAS* (communal centers for social action).

The logic of decompartmentalization is thus a necessary dynamic with which to build the basis for coordination between the different professionals. On the ground, it is a question of articulating the interventions of each one while advocating in the initial stage the dialogue and lastly the performance. At the level of financiers, it is a question of relying on local dynamics to tend towards *appropriate remedies* and *efficiency* (Ministère des affaires sociales et de la Santé 2013). The final objective of these two poles of health professionals is therefore not the same. However, the initial need is of the same nature, especially since performance and efficiency are not equivalent: the distortion between what is played out on the ground at the professional level and what is expected in corollary at the regulatory level is the great challenge that has faced the French Government for more than 20 years.

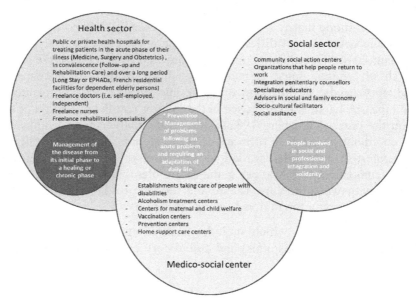

Figure 2.1. *Professional poles that collaborate for health*

1 An equivalent of a *département* in the US would be a county.

2.1. Circumscribing the notions of *coordination, pathways* and *complexity*

To meet the requirements of national health policy guidelines, the *Haute autorité de santé* (French national authority for health) proposed to envisage coordination as a "deliberate organization of patient care activities between several professionals in the health system" (Haute autorité de santé 2014). A corollary to this wish: coordination implies that the professionals communicate with each other, which translates into *the exchange of information between the participants responsible for the different aspects of care* (Haute autorité de santé 2014). The French government has thus launched various experiments, the most recent of which is the *Territoire de soins numériques* (digital healthcare area), which is supposed to make the dematerialized sharing of health information a key tool for the coordination of pathways, particularly when it comes to complex situations (Loi n° 2004-810 du 13 août 2004). With its digital tools, e-health can only promote both superficial exchanges and consultations: a proactive evolution of positions is necessary and must be shared by professionals who are often distant from each other in terms of methods, concepts, financing, legal structuring and objectives. Exchange, information sharing and consultation *on a proactive basis* therefore represent the first organic basis inherent to the notion of coordination.

In order to make the simultaneous action of several effectors coherent and synergistic, a structuring nature is needed to form a second basis of coordination. It must call for the *formalization* of exchange, sharing and consultation processes, of which the digital tool is only a carrier. However, it only takes a bit of effort to rely on a structuring approach for efficiency purposes in the strict sense. Historically, the *pathway of coordinated care* has been a first step implemented by *l'Assurance Maladie* (French health insurance), framing the use of healthcare with "provisions relating to the organization of the supply of care and the medicalized control of health expenditure" (Loi n° 2004-810 du 13 août 2004). This pathway of care was then posted as part of the reinforcement of the role of general practitioners to "encourage patients to seek one of the professionals providing primary care as a priority, who must then ensure the referral of patients to care providers and medico-social services" (Loi n° 2004-810 du 13 août 2004). It was only in a second phase that the notions of the *health and life course pathways* appeared, in order not to overburden the necessary individualization of patient management and to associate it with a consideration of the social situation (Bloch and Henaut 2014). The *Haut conseil pour l'avenir de l'assurance maladie* (French high council for the future of health insurance) has thus introduced the notion of health pathways in search of a quality of care and a social approach, thus comprehensive, particularly with regard to the difficulties encountered by people with **chronic pathologies** (Haut conseil pour l'avenir de l'assurance maladie 2012). However, the "pathway" dynamic actually has an objective for regulatory authorities, the *appropriate remedy*: "To ensure that a population or individual receives the right care by the right professionals in the right structures at the right time and at the best cost" (Agence régionale de santé Loire Rhone Alpes 2013). To this end, although idealistic to say the least, it is necessary to both personalize the pathway with regard to the uniqueness of each individual and to protocolize it by virtue of equity, but above all with rationality: although a person's health pathway is unique, it is necessary to identify the types of pathways present in the population and to calibrate and anticipate the resources necessary for their good management (Agence régionale de santé Loire Rhone Alpes 2013). It is easy to recall what has altered the logic of the pathways since its accession to the rank of national priority, when two antinomic approaches become telescopic: "one is derived from good practices and clinical standards but gives little room for the individual desires and rhythms, while the other takes into account all the specificities and individual desires but is more resource-consuming" (Bloch and Henaut 2014).

DEFINITION.– "**Chronic diseases**": diseases requiring management over a period of several months or several years and some of which last an entire lifetime. These are mainly diabetes, renal failure, cancer, Alzheimer's disease, neurological or degenerative diseases (myopathies, Parkinson's disease), rare or inflammatory diseases (cystic fibrosis, rheumatoid arthritis), long-term mental disorders, AIDS and hepatitis C.

If most people suffering from chronic pathologies or with a **loss of autonomy** are today likely to benefit from the organizational logic of health pathways according to the common modes of practice of professionals, there are situations, called *complex*, where both participants and patients are in difficulty. According to the consecrated definition, these are "situations calling for a diversity of health, medico-social and social workers, to whom the primary care physician cannot respond with their own means" (Direction générale de l'offre de soins 2012). The intention here is not to focus the perception of complexity on a criterion of age, a given pathology or a handicap, or to favor a pragmatic approach in the practitioner, with the corollary of directing them towards resources able or dedicated to helping them in their practice, such as *support systems for primary care*. Complex situations often involve people with chronic pathologies, loss of autonomy, disabilities or psychopathological disorders, or those in precarious situations. This is what is clearly meant when the *Haute autorité de santé* proposed five major dimensions of complexity that could serve as identification criteria (Haute autorité de santé (HAS) 2014).

DEFINITION.– "**Loss of autonomy**": the inability of a person to perform certain acts of daily living on his or her own. It is sometimes called dependence. It varies in intensity and nature, depending on the individual, and over time: for example, some people will be said to have lost their autonomy because they can no longer stand up on their own, while for others, the loss of autonomy will manifest itself in memory difficulties.

Criteria for identifying health complexity:

1) *physical* health (chronic diseases, polypathologies, comorbidities, cognitive decline, unstable pathology, difficult diagnosis);

2) *mental health* (psychiatric pathology, psychological distress, addiction, refusal of care);

3) *demographic characteristics* (advanced age, low level of education, language barrier);

4) *social capital* (poverty, lack of social insurance, caregiver burnout, lack of family or social support, isolation, homelessness);

5) *health and social identification experience* (poor quality of life, repetitive hospitalizations, readmission to emergency, difficult orientation in the health system, limited therapeutic compliance).

Box 2.1. *Criteria for identifying health complexity*

Complexity is in fact relative and subjective in character: "complexity can appear to the person who lives it or who evaluates it or tries to understand it, to each of the articulations of this definition: in the form or type of limitation or restriction, in the articulation with environments, in the combination of alterations, or in the overall dynamics of all these elements that the person experiences and/or understands with difficulty or not at all"

(Centre d'études, de documentation et d'information de l'action sociale 2013). Thus, recognizing a complex situation may call for different types of approaches, possibly combined, depending on who studies, confronts or evaluates it: on the basis of specific criteria, according to the judgment of the health professional (knowledge, experience, clinical sense) or according to predictive modeling based on a series of indicators.

DEFINITION.– "**Complexity**": a complex health system is a group of individual agents who have the freedom to act in ways that are not always predictable and whose actions are interconnected. Complexity is close to chaos.

As the initial actors within the coordinated healthcare pathways, people, including their entourage who help them, find it difficult to find a clear way through the offered healthcare and services. It is complicated for them because it is fragmented, compartmentalized and lacking in apparent coherence. Faced with the constellation of institutions and professionals, with by whom they are sometimes more confronted than helped, users find it all the more difficult to live with the complexity of their health situation because it is partly linked to the administrative complexities generated by the diversity of the status, prerogatives and competences of all the components of the system. The notion of course is thus once again under the seal of distortion, charged with "an *a priori* humanist and even individualist, who counterbalances their current formal managerial use" (Chauvière 2010). On the one hand for healthcare users regarding the complexity of the system, and on the other hand for the regulators of this system, which must be rational but is still confronted with the uncertainty often linked to human nature, it is the real face-to-face meeting that characterizes many health pathways and calls into question attempts at management or anticipated organization.

Various terms and notions often used in the logic of the pathways translate these difficulties well: to tend towards the *anticipation* and the *avoidance of ruptures*, to release the *brakes*, to ensure *fluidity*, etc. A holistic but realistic definition of the pathway and its attributes was proposed in Michel Chauvière's sociological approach:

> "The pathway is generally presented as ensuring not only the concrete continuity of a singular path towards a solution, a realization in relation to recognized individual needs; but also, that it must allow for better mobilization against discontinuities

and other ruptures of all origins (technical, administrative, [...] or simply relating to ordinary life and its hazards..." (Chauvière 2010).

The sociologist also sets the spirit of their discourse in opposition to the French context: "if there are obstacles in this perspective, they would be first of all ontological, in the case of the human being as such, and political, in the case of the government of humans as humans and by humans. At the ontological level, the issue is the individualization accelerated by the very idea of reducing the human being to their journey and, at the political level, it is the decline of the idea of society or collective, and also of rights claims and obligations of intervention of powers for a common destiny" (Chauvière 2010). In the same vein, the sociologist summarizes health assistance through their own system, necessarily institutionalized: "... whatever one has imagined or fantasized about behind pathways, devices, and coordination (these three terms visibly making a system), alongside rights (subjective of course but also positive), there is always more of the institution, that is to say global, integrated, living, and sustainable supports (thus hospitals, schools, establishments, employment centers, Social Security organizations, etc.) for people, and especially the most deprived of personal resources" (Chauvière 2014). In this vision, the free world is not excluded, but its place is not defined. At the time of the "ambulatory turn", the stakes are thus multiple and remain marked by the seal of the complexity of the system.

2.2. Identifying the structural elements of pathway coordination

In an attempt to emerge from the contradictions noted above, several structural and significant elements seem necessary at the organizational and operational levels. Experiences from health network practice are valuable in this respect. Historically, many successful collective initiatives, leading to a genuine decompartmentalization between professionals, of a bottom-up nature, have been conceived in the last 20 years from the perception by the professionals themselves, in the field, of such a need: certain services dedicated to disability stemming from associations created solely on the basis of the motivation of families, multidisciplinary evaluation and care structures including those dedicated to chronic pain, but above all *health networks*. It should be noted that legislators and public authorities often drew on the experiences of health networks and tried to shape them in return regarding their strategic objectives, ultimately asking those very professionals who had inspired them to comply with new rules and specifications or even to be replaced by someone coming directly from a top-down organization.

It is possible to identify the basic determinants of coordination based on the experience of health networks, in light of the findings previously presented:

Basic determinants of coordination:

1) *define who is the referee of the pathway*, the person or professional ensuring the follow-up of the pathway in all its dimensions and towards whom all the others converge the data and can turn to; they have an overall and updated vision of the situation (*a priori*, it is the attending physician, but there are situations where this is not possible);

2) *assume subsidiarity between professionals*, while some consider it as a dependency on others;

3) *abandon corporate positions*, ways of doing and being, reflecting a group, a structure or a method with which any professional identifies, sometimes obstinately.

Box 2.2.

All in all, it is **cooperation** between professionals which is the key term and which is hidden behind that of coordination, a collective human dynamic which makes it possible to accomplish things together as a tightknit unit, that which we build together ourselves and not that which is provided by a guardian: from proximity, when patient and care provider cooperate under the sign of the **therapeutic alliance** (Collot 2011; Suarez Herrera 2013), to the financer who, like the others, must be a partner of the collective of professionals (on behalf of the "community"), when they accept the **participatory evaluation** game (Mick 2016).

2.3. Mobilizing together

The French Government's efforts in recent years to optimize healthcare pathways, and consequently their coordination, have mainly focused on the elderly or those at risk of loss of autonomy, as well as on so-called polypathological patients (many of whom are also elderly). With the shift to ambulatory care and the concern to reduce hospitalization and promote return and home care, the consolidation of certain organizations has been unsuccessful because it has been of interest to offer services located in hospitals for the most part, including territorial health networks, located at the crossroads between the independent and hospital worlds and the health and social sectors. However, it is not unrealistic to imagine an initial structuring of the multi-professional and overall approach of a person in proximity, in the "gap", in connection with primary care teams as already organized by certain professionals in the area.

The real problem of the coordination of a pathway is not so much at the level of the initial evaluation of the situation, particularly complex, or even if it requires a certain amount of expertise (Gillet 2016): it is at the level of the implementation of the care itself and its follow-up. The frequency of chronic pathologies and the intertwining of bio-psycho-social determinants underlying the complexity of situations, and consequently the need to use various specific techniques in parallel, encourage the recognition of a **pluriprofessional team** for such an implementation. It is the initial consultation time between the main professionals around the person, including the primary care team in the first place, which is the key element of coordination, and especially the starting point for the "setting to music" of a difficult score, with "musicians" scattered around the "conductor", the point of contact for the pathway (Haute autorité de santé 2008). The attending physician intervenes here, in all cases, as the pivot of coordination in their overall approach, although the coordination of the pathway of a person with a chronic pathology, disability or loss of autonomy may require as such a knowledge of protocols, methods and tools, as well as professionals implementing them within an area and then registering another point of contact in close link with them: coordination of support for health networks, case management of MAIA (*Méthode d'action pour l'intégration des services d'aide et de soins dans le champ de l'autonomie* (method of action for the integration of assistance and care services in the field of autonomy) devices. The AFM, *Association française contre les myopathies* (French association against myopathies) such as *Téléthon en Auvergne-Rhône-Alpes*, for example, is the third level of local information and coordination centers.

Like these primary care support systems deployed on an outpatient basis, a hospital team can also provide support and resources, such as an interface system that must provide several services, both assessment and care as needed, in direct and permanent contact with the outpatient world (for example, teams dedicated to chronic pain). It is the coherence of the actions of one and the other which is therefore the key, downstream of sharing and exchanges, and of consultation. Why is this simple logic not the seed of coordination between health professionals in everyday reality?

DEFINITION.– "**Primary health care**": according to the World Health Organization, "primary health care is essential health care based on practical, scientifically viable, and socially acceptable methods and technology, made universally accessible to individuals and families in the community through their full participation and at a cost that the community and the country can bear at each stage of their development in a spirit of self-responsibility and self-determination".

2.4. Coordinate around the person

On the basis of experiences from the area's multi-purpose health networks, it is legitimate to propose a deployment of coordination based on several successively nested cornerstones, which serve as authentic benchmarks for the health pathway.

– FIRST CORNERSTONE: this is located at the same time of the definition of the person's current problems by their attending physician. Ideally, all practitioners have assessment tools that allow them to keep a file according to a form of organization adapted to the multidimensionality of the individual situation. However, in addition to motivation, it takes time for the practitioner to come to grips with it and to receive all useful information, including social information.

– SECOND CORNERSTONE: this concerns the sharing and exchange of the objectives of care defined with the person and for all of the professionals. If an information system shared by the professionals is indeed the appropriate tool when the patient has authorized/empowered them, for the time being, this sharing still remains in written form (dematerialized or not) or telephone-based (Haute autorité de santé 2008). Even today, electronic messaging, which reduces delay and hardware archiving, is most of the time unsecured, inducing the involuntary provision of real "big data" for future providers of multiple paid services for both health and consumption. In fact, for the time being, it is not a question of sharing or exchange, but of making information available (Haute autorité de santé 2008).

– THIRD CORNERSTONE: this concerns the coherence of diagnostic positions (in the broad sense of the diagnosis of the situation) between professionals. First, any disagreement between one or more professionals about a "causal" diagnosis deserves exchange and consultation, right from the start, that is, before implementation of management objectives. Carving such a stone is almost an ancient myth, in a society where the (human) professionals are (fortunately) free to think (of a diagnosis, a cause) but consolidate their narcissism (obligatory) and professional identity (knowledge related to the profession, specialized expertise, corporatist postures) by considering without questioning or simple clinical re-reading of their diagnosis as good and delivering it as a "contribution" to the person and to others. Secondly, if we focus on the bio-psycho-social approach, there is not only a diagnosis, but there are problems (Haute autorité de santé 2008; Boureau 1994): it is these multifactorial determinants that deserve an evaluation by and therefore with the assistance of each professional, based on their expertise and experience, to contribute to collectively constructing a picture of the situation, around the attending physician and the patient's advocate if necessary. Consultations between stakeholders act as the glue for ideas, of which we know all too well the parasitic constituents who prohibit entry for regulars as well as others, with dedicated/paid time and proactive positions (Haute Autorité de santé 2008). Indeed, this consultation, particularly when the professionals do not know each other, requires time and motivation, and when it can take place in a pluriprofessional way and on the scene (ideal and difficult to achieve in practice but today required by the government as an "open sesame"), it is carried out with a time frame that is perplexing with respect to the notion of efficiency.

– FOURTH CORNERSTONE: this concerns the *coherence of the positions of care* between various professionals. If there is concertation around the clinical picture and its etiological factors on the one hand (care project), and the need for help on the other hand (help and support plan), the "road map" of care or *personalized health plan*, elaborated with all of the data coming from the involved professionals by the attending physician and/or the identified point of contact, proposes interventions to the different professionals around the person and this in subsidiarity, as the construction of the term "coordinate" elicits: a healthcare professional requests others to carry out an act or a service identified as appropriate or necessary. In addition to having a natural propensity in an independent practice, the caregiving world in general is reluctant to follow the instructions of a colleague unless it is a peer or a companion, or after real consultation and consensus, and even more so when it is a health professional from another health sector. The gap is even greater between professionals coming from a combination of health and non-health spheres, who do not speak the same language. We then find the multiplicity of acts and services superimposed without coherence, proposed in bulk to a person, appended to the various diagnoses and needs identified by the various "experts" requested by them or a doctor or another health professional.

– FIFTH CORNERSTONE: this is the chronological relationship between the various interventions identified during the declination of the objectives of care and their implementation in a joint way by the professionals around the patient and the patient themselves, this under the benevolent eye of the attending physician. The differential temporality at the level of each intervener, from medical to social (appointment times and schedules, need for successive sessions, proximity to home, availability and means for the patient to go to them), is one of the reasons for the daily headache faced by patients and healthcare professionals. Deploying the various objectives on an outpatient basis in a coherent and effective manner is a long process for the patient and the attending physician. It is also at this level that the support of this team can return to dedicated devices, such as health networks, or that of an *assessor* when possible, or to a primary care team coordination nurse, such as those resulting from the ASALEE (*Action de santé liberale en équipe*) device (independent healthcare team actions) (Ministère des affaires sociales, de la santé et des droits des femmes 2015).

Finally, it is necessary to add scaffolding around these cornerstones, those things which are *unconventional*, *not academic*, and *not professional*. Men and women of various arts, very often not health professionals (note: *psychologists are for us health professionals*), diagnose, prescribe, affirm and build parallel or even contradictory pathways of any other realm, or substances and other products that are active (therefore potentially dangerous) or not, used in a hidden way most of the time in parallel with the treatments prescribed by healthcare professionals (Académie nationale de médecine 2013; Testard Vaillant 2014). The people around the patient also have and always do offer their contribution: spouses, caregivers, family members, neighbors, friends develop their vision of the causes of poor health and related problems, and suggest their often "radical" solution, stemming from their own experience as a patient, from the Internet or from "someone who knows someone who has already had...". There is no question of coordination here, whereas for many patients, these naturally invited and inevitable influences have a major role.

2.5. An interpretation of reality

QUESTION.– What is happening today with people regarding their health pathway, which should be a matter of coordination between professionals but is more like a market for services at the mercy of free trade?

2.5.1. *Lack of sharing and exchanges between professionals*

When a treating physician refers a patient to another professional, their referral can be disconcertingly brief, simply expressing the need for advice and/or care: everything depends on the role and responsibility assigned by the treating physician at the time of referral. In addition, how many health professionals send copies of their findings, recommendations and prescriptions (consultation reports) to the other professionals that request them, such as a physiotherapist who would need them to narrow their physiological diagnosis (regulatory necessity), their medical diagnostic opinion (natural tendency) or their prescribed and non-prescribed therapeutic approaches? On the other hand, do physiotherapists or psychologists, whether solicited or not, or alternative practitioners, turn to the attending physician to inform him/her of their findings, opinions or perhaps therapeutic choices? Among the autonomous doctors in charge of examining an application for disability or allowance, are the medical officer of the *Caisse primaire d'assurance maladie* (CPAM), who must make a decision with serious social consequences; the occupational doctor, who must decide on the suitability for a post and possible adaptation or reclassification; and finally the medical expert, who assesses the damage, levels of disability and rehabilitations. Which of them has an overall reading, displaying the keys to coordinated management proposed by the attending and/or referring physician, rather than a simple medical certificate, to make the right decisions in an informed manner? In daily practice, can we envisage a multi-professional consultation meeting between all these doctors and all these professionals, or should we spend time taking out a few opinions rather than positions and draw as best as we can a meaningful and consensual roadmap for care?

2.5.2. *Versatility and overlay of "diagnostics"*

Besides the fact that a "causal diagnosis" will at some point be made by a doctor, there will be among their colleagues those who do not wish to hear about it and have a completely different diagnosis, those who reduce a patient's situation to psychological suffering, or those who attribute it to the misdeeds of modern society (vaccination, endocrine disrupters, heavy metals, malignant microbes (Lyme disease), physiological hypersensitivity or allergies (milking, gluten, meat), electromagnetic waves, redemptive (spiritual) influences. In general, a first diagnosis mentioned by the attending physician is based on (1) the diagnosis made by a specialist and not necessarily shared with the attending physician; (2) the diagnosis made during care by a non-physician; (3) the diagnosis that is necessarily different and incomprehensible by others when delivered by an alternative provider; (4) the diagnosis discussed by non-healthcare professionals, partly biased by their own experience and representation but also by their professional objectives; (5) those of their (many) family and friends. The patient has at least one potential compass to find their way in this myriad of points of view: their attending physician, if they know him/her well enough and the physician assumes the place as the pivot of the health pathway, and if they have the necessary information. When this is not the case, the work of proofreading and checking consistency can possibly be ensured by an identified point of contact, doctor or not, if they also have the necessary information available.

2.5.3. *Telescoping medical positions*

Every health professional has knowledge and experience, but they also carry out tasks in a given field of expertise and prerogatives. Thus, in the field of healthcare, a doctor prescribes and paramedical workers perform, while in the medico-social and medico-professional field *(CPAM, Maison départementale pour personnes handicapées (MDPH), médecine du travail)*, a doctor rules and sets up a regulated framework which classifies the patient with regard to their aptitudes and rights. It is rare that these different medical professionals follow the same roadmap, with different objectives and constraints. Thus, very often, and although they come from the same (health) environment, some work in care and only observe the evolution of the patient's social identity without contributing to it, and others work and shape this identity by observing care without taking part in it. Fortunately, in the world of healthcare, it is common for the attending physician and paramedical worker (nurses, physiotherapists, pharmacists, psychologists or others, particularly in the field of psychocorporal approaches) to spontaneously combine their care with benevolent listening and support of the patient. Yet, do we know what happens in this space that is verbal and not transcribed, outside of the care itself, and which is so crucial to helping the patient on their health pathway?

Difficulties between the various actors

Generally speaking, there are four kinds of difficulties (Rainhorn 2011; Bruyère 2008):

– **structural difficulties**, which are linked to the telescoping of the organization of the various devices and structures whose teams meet in the field, with multiple constraints and a very different temporality, which impose times and spaces of meetings and exchanges to tend towards efficiency and coherence;

– **cultural difficulties**, which are linked to the different representations between professionals of their role and that of others, with which crystallize (1) fears linked to the pooling of information, overlapping of competences, shifting of tasks, dilution of competences, "flight" of patients and the regard of others on their profession; (2) divisions (corporate, in particular) and (3) attitudes (paternalistic vs. constructivist);

– **behavioral difficulties**, which are linked to the context, most of the time to the liabilities and the local history of the professionals between them;

– **organizational difficulties**, which are linked to the modes of organization and functioning of each participant (team/individual, hospital / independent practice, salaried/contractor, hourly salary / fee for service).

One of the challenges of successful pathway coordination is to identify problems in these four areas and propose solutions, using imagination, flexibility and adaptability. In an approach promoting meetings between multiple stakeholders, it will be a matter of creating a formal, consensual and meaningful content: reports, personalized health plans, recommendations, organization of stakeholders and interventions, and evaluations. In short, seeing different points of view to achieve a common objective: "comprehensive and co-produced care that improves the quality of healthcare" (Bruyère 2008). The structured organization of a system is therefore not sufficient for coordination between its professionals to be organized: the proactive character, the effective mobilization, the renunciation of one's previous position and the acceptance of the opinions of others are the keys that remind us that it is the place of humankind at stake.

2.6. References

Académie nationale de médecine (2013). THÉRAPIES COMPLÉMENTAIRES : Leur place parmi les ressources de soins. Report. Available: http://www. academie-medecine.fr/wp-content/uploads/2013/07/4.rapport-Th%C3%A9rapies-compl%C3%A9mentaires1.pdf.

Agence régionale de santé de Rhone-Alpes (2013). Parcours de soins, parcours de santé, parcours de vie. Available: http://ars.sante.fr/Parcours-de-soins-parcours-de.148927.0.html.

Bloch, M.-A. and Henaut, L. (2014). *Coordination et parcours. La dynamique du monde sanitaire, social et médico-social.* Dunod, Paris.

Boureau, F., Doubrère, J.F., and Luu, M. (1994). Approche clinique des patients douloureux chroniques. *Rev Prat.*, 15, 1880–1885.

Bruyère, C. (2008). Les réseaux de santé français : De la compréhension d'une nouvelle forme d'organisation des soins à la construction d'un modèle de management paradoxal. Université de la Méditerranée, Aix Marseille II.

Centre d'études, de documentation et d'information de l'action sociale (2013). Les situations de handicap complexe. Besoins, attentes et modes d'accompagnement des personnes avec altération des capacités de décision et d'action dans les actes essentiels de la vie quotidienne. Report. Available: http://www.creai-idf. org/sites/cedias.org/files/rapport_public_les_situations_de_handicap_complexe_ cedias_clapeaha_cnsa_sectio n_economie_sociale_chorum_juin_2.pdf.

Chauvière, M. (2010). *Trop de gestion tue le social.* La Découverte, Paris.

Chauvière, M. (2014). La fluidité des parcours et les freins. Available: http://www. chemea.fr/images/banque/files/Michel_CHAUVIERE.pdf.

Collot, E. (2011). *L'alliance thérapeutique*. Dunod, Paris.

Direction générale de l'offre de soins (2012). Guide méthodologique pour l'évolution des réseaux de santé. Available: http://social-sante.gouv.fr/IMG/pdf/Guide_reseaux_de_sante.pdf.

Gillet, D., Heritier, S., Garcia-Porra, C., Varigas, M., Bochet, E., Ramponneau, J.-P., and Mick, G. (2016). La concertation pluriprofessionnelle au service du parcours de santé du patient douloureux chronique. *Douleur et Analgésie*, 29, 158–162.

Haut conseil pour l'avenir de l'assurance-maladie (2012). Avenir de l'assurance maladie : Les options du HCAAM. Available: http://www.securite-sociale.fr/IMG/pdf/l_avenir_de_l_assurance_maladie_les_options_du_hcaam.pdf.

Haute autorité de santé (2008). Douleur chronique : Reconnaître le syndrome douloureux chronique, l'évaluer et orienter le patient. Available: http://www.has-sante.fr/portail/upload/docs/application/pdf/2009-01/douleur_chronique_recommandations.pdf.

Haute autorité de santé (2014). Note méthodologique et de synthèse documentaire. Coordination des parcours. Comment organiser l'appui aux professionnels de soins primaires ? Available: http://social-sante.gouv.fr/IMG/pdf/Guide_reseaux_de_sante.pdf.

Loi n° 2004-810 du 13 août 2004 relative à l'assurance maladie [online]. *Journal officiel*, (190). Available: https ://www.legifrance.gouv.fr/affichTexte.do?cidTexte=JORFTEXT000000625158&categorieLien=id.

Mick, G. and Moyenin, C. (2016). L'évaluation initiale d'une plainte douloureuse chronique : Réflexions et propositions pour un guide pratique. *Douleur et Analgésie*, 29, 130–140.

Ministère des affaires sociales et de la santé (2013). Stratégie Nationale de Santé. Available: http://travail-emploi.gouv.fr/IMG/pdf/1_Strategie_nationale_de_sante_ce_qu_il_faut_retenir-2.pdf.

Ministère des affaires sociales, de la santé et des droits des femmes (2015). Modalités de mise en œuvre locale du protocole de coopération médecin/infirmière porté par l'association ASALEE. Available: http://circulaires.legifrance.gouv.fr/pdf/2015/05/cir_39573.pdf.

Mintzberg, H. (1982). *Structure et dynamique des organizations*. Editions d'Organisation, Paris.

Rainhorn, J.D. and Burnier, M.-J. (2011). *La santé au risque du marché : Incertitudes à l'aube du XXIe siècle*. PUF, Paris.

Suarez Herrera, J.C. and Grenier, C. (2013). L'évaluation participative : Levier de changement pour l'amélioration de la performance du système de santé ? Université d'été de l'ANAP. Available: www.performance-en-sante.fr/fileadmin/5.SUAREZ-GRENIER.pdf.

Testard Vaillant, P. (2014). Médecines alternatives. *Science et Santé*, 20, 22–33.

Part 2

Optimization of Flows within a Hospital

Introduction to Part 2

According to certain brainwaves, hospitals can be compared to a company or an anthill.

The economic and organizational models of hospitals have considerably evolved over decades, constrained on the one hand by the increase in knowledge, the emergence of new diseases, the development of technologies, population growth and the aging population. An old-fashioned nurse, very devoted to her patients under the benevolence from high above, made room for a plethora of people qualified as doctors, occupational therapists, psychologists, physiotherapists, social assistants and administrative agents, organized according to a hierarchy of well-defined disciplines in a brand new hospital where the elevators direct patient needs. The patients meet the doctors, the nurses meet the pharmacists who themselves meet the doctors under the control of the computer scientist listening to the nurse on their computerized care file, and the patient receives a message on their smartphone, notifying them of their next appointment when they are not even sure they will even survive their imminent surgery.

Removing all the stains, relieving suffering, gardening, accepting what they could for gifts or fruits of their labor, the old-fashioned nurses dedicated their lives to God but especially to their hospital, concluding most often the stays of the sick by closing the wooden lid leading towards eternity. With the advent of vaccines, the long battle of another woman, Marie Curie, and antibiotics, health costs and prospects have dramatically changed. Healthcare funding therefore had to be rethought to respond to this revolution. Already, the Second Empire under the impetus of industrial development had designed a system of financing patients launching the beginnings of the French government health insurance. The Second World War ushered healthcare into an era of massive funding, and hospitals were able to organize themselves. In 2004, this financing system led to the implementation by the French government of the "activity-based pricing" system. In other words, just like a hammer or a dress, a patient's stay becomes a product invoiced to the client: *Caisse d'assurance maladie, CNAM* (French national health insurance fund). The hospital has thus transformed itself into a health manufacturer with projects, translated into strategy, policies and action plans. Its purpose is to produce and provide care and services for a group of "customers" or "users", by achieving a balance of its expenses and income accounts. The "manager" decides what services the caregivers perform, and the patient complains! All this in an anthology of exchanges, data transfers and collaborations, including a multitude of information that everyone must appropriate to best take care of the patient!

Nevertheless, behind this "industrialization of illness", the hospital remains a vast field where everything must be done to enable the structure to perform its function as well as possible, despite the money predators. Working relentlessly according to a well-defined hierarchy, a multitude of ants arrange, feed and repair each day of the year according to a network organization adapted to the vicissitudes of the environment. Illness is the object of their labor, the patient the objective of their days and even if the great kick of finance temporarily disrupts its functioning, all the worker ants resume the path of their mission, abandoning neither the patient nor the illness. They recreate a space of care in the awareness of external hazards and re-adapt their structure to the context of the stock that welcomes them. They take the food that comes to them and organize it according to the needs and skills of each one, creating a real network capable of conditioning the meaning of their lives as well as possible.

Professions are changing and techniques are adapting. Doctors join forces with engineers who put their individual skills at the service of a collective intelligence, juggling economy and efficiency to ensure the best for patients in a world with strong CAC 40 (a benchmark French stock market index) accents. The management of flows, whatever they may be, is part of this reflection. The optimization of networked processes requires the control of flows, which will be shown in this chapter.

3

Decision Support Methods for Efficient Flow Management in Medical Device Sterilization Departments

3.1. Context and motivation

The primary objective of healthcare provision is to provide quality care to its patients. It should be noted that France ranks 4th among the Organization for Economic Co-operation and Development (OECD) countries for its level of health-related expenditure (11% of GDP in 2016, against 3.4% in 1960 and 6.3% in 1980), behind the United States (17.2% of GDP in 2016), Switzerland (12.4% of GDP in 2016) and Germany (11.3% of GDP in 2016). Major reforms have therefore been undertaken in France to control health expenditure, without degrading the quality of care. At the same time, hospitals are facing changes in the behavior of patients (who are becoming more demanding) and medical staff (whose recruitment and retention are more difficult due to medical shortages). It is a system dominated by supply shifting to a patient-centered system, with strong resource constraints and increased demand. The healthcare production system must therefore now not only meet patients' expectations by providing quality care but must also be efficient in terms of costs and deadlines. French hospitals are now to be evaluated (by various authorities) according to criteria falling within the Project Management Triangle, also known as the time-cost-quality triangle, which is well-known in the industrial sector. The notions of waiting times, resource utilization rates, respect for working time and risk management, are now at the center of hospitals' concerns.

Chapter written by Maria Di Mascolo.

DEFINITION.– The OECD has a mission to promote policies that will improve economic and social well-being around the world. This was created 50 years ago (1948) following the Second World War to administer the Marshall Plan in Europe, financed by the United States. Strictly European at first, it became global with the annexation of the United States and Canada in 1960 followed by Japan in 1964. Today, 35 countries are members.

In this context, it is therefore essential to have available scientific methods, derived from methods from the industrial sector, for the design and management of healthcare production systems to have the most efficient organization possible. The major difficulties for the development of such methods are the complexity and diversity of healthcare production system organizations, the importance of the human factor and uncertainties, and the need for an effective coordination.

The work that has been carried out focuses on improving the organization of healthcare production systems, in particular, by identifying "good organizational practices" to satisfy the various stakeholders of these systems (patients, medical and paramedical personnel, society), through the achievement of performance indicators such as the effectiveness of interventions, the quality of care from the patient's point of view (for example, adherence to visiting hours), the costs associated with operating the system and the quality of work for staff (for example, securing working hours, minimizing stress at work, or taking account of caregivers' personal preferences), or robustness in the face of uncertainties and risks.

Decision support methods and tools have been developed to guarantee good performance for healthcare production systems, despite uncertainties. For that, an important step is the modeling of the studied system, which allows us to represent and understand the studied system. Methods and tools to predict system performance and to assist decision-making then complemented this approach.

This work was carried out in collaboration with A. Gouin of GIPSA Lab, E. Marcon of LASPI and then DISP, M.-L. Espinouse, A. Sebö and J.M. Flaus of G-SCOP, L. Schwob of CHPSM, notably as part of the theses of K. Ngo Cong (2009), Ozturk (2012) and Negrichi (2015), and the 2E2S Projects

(2008) – the "Electronic sterilization departments survey", and "OptiCLS" (2011) – and Ozturk's (2011) "Optimisation du chargement des laveurs dans un service de stérilisation hospitalière" (Optimization of washer loading in a hospital sterilization department).

3.2. Sterilization departments for medical devices

Many medical treatments require the use of instruments called medical devices (MDs).

DEFINITION.– A medical device is any instrument, apparatus, equipment, material or product (except products of human origin, or other items used alone or in combination, including accessories and software necessary for its proper functioning) intended to be used in humans for medical purposes and whose principal intended action is not obtained by pharmacological or immunological means or metabolism, but whose function may be assisted by such means. A medical device is also software intended to be specifically used for diagnostic or therapeutic purposes. Medical devices are classified into four categories according to their potential health risk. Specific evaluation and control rules are associated with each category:

– *Class I (lowest hazard class)*, which includes, for example, corrective eyewear, vehicles for the disabled, crutches, etc.;

– *Class IIa (potential moderate/measured risk)*, which includes, for example, contact lenses, ultrasound equipment, dental crowns, etc.;

– *Class IIb (potential high/significant risk)*, including condoms, lens disinfection products, etc.;

– *Class III (highest risk class)*, which includes, for example, breast implants, stents, hip prostheses, etc.

Some MDs can be reusable and must therefore be sterilized between uses. In hospitals, specific departments are devoted to this task: these are the sterilization departments.

The sterilization of reusable medical devices removes microorganisms from them. It is an essential action to fight against infections. The sterility of a medical device is guaranteed by compliance with the performance of a set of operations necessary to obtain and maintain the sterile state of this device (AFNOR standard). Reusable medical devices are instruments that will undergo multiple sterilization processes. The production process of sterile medical devices is therefore represented by a cyclical pattern, called a sterilization loop. This loop comprises several stages, namely use, pre-disinfection, rinsing, washing, packaging, sterilization itself, quality control of the various operations and storage. These different steps are separated by more or less substantial transfer phases and by intermediate storage phases. Figure 3.1 illustrates the different steps in sterilizing an MD.

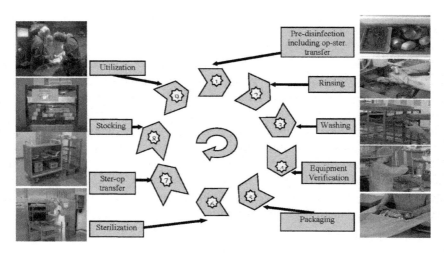

Figure 3.1. *Sterilization loop (Ngo Cong 2009)*

This sterilization activity is both costly and strategic, because it is the basis for the operation of operating theaters and other hospital services.

In the literature, most of the articles dealing with the sterilization process propose technical studies on the rules to be complied with to ensure the sterility of medical devices (Rutala 2004; Smyth 1999; McNally 2001). In the sterilization departments we visited, particular attention is paid to compliance with these rules, all of which are based on ministerial recommendations (Ministère de l'emploi et de la solidarité 2001). On the other hand, the organization of production varies from one establishment to another and, often, the flows are managed without precise rules (apart from minimum rules imposing a directed flow from dirty areas to clean areas), nor concern for optimization. Even if the operation of the sterilization departments we visited seems generally satisfactory to their managers, after more in-depth discussions, we realize that deficiencies exist (we can cite, for example, containers of MDs delivered late, containers forgotten in storage or a very irregular workload). It should be noted that the sterilization department is both a client and a supplier of the operating theaters and does not control the arrival of MDs from the operating theaters.

The work we are carrying out, and which we are presenting here, precisely concerns the study and improvement of the organization of the production of sterile medical devices in hospitals and, more recently, risk analysis. We started from particular cases, to go towards generic models, and we started from performance evaluation to go towards a coupling with optimization and risk analysis.

3.3. Our contributions

3.3.1. *Field study and performance evaluation of a sterilization department*

We present here solutions to help manage a centralized sterilization department and we illustrate the contribution of our proposals on a real case: the sterilization department of the Centre Hospitalier Privé Saint-Martin (CHPSM) (Saint-Martin Private Hospital Center). A production flow simulation study for this sterilization department was carried out and highlighted certain deficiencies linked to the management of the department and demonstrated possible improvements. We were particularly interested in streamlining production and reducing retention time, either by modifying flow creation (action on pre-disinfection time) or by modifying sterilization department management (modification of personnel planning).

This work was carried out in the thesis of Ngo Cong (2009) and was the subject of (Ngo Cong 2007) and (Di Mascolo and Gouin 2013), which provides details of the study and the results obtained, which are summarized here.

We started our work with a field study at CHPSM, which allowed us to build the model of the sterilization process shown in Figure 3.2, which includes several steps after the use of MDs. This use corresponds to the 14 surgical rooms present at CHPSM.

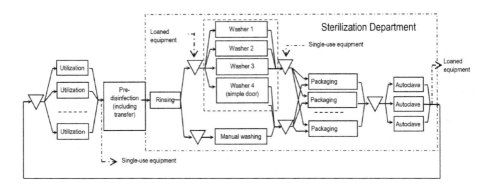

Figure 3.2. *Organization of the sterilization process at the CHPSM*

These steps follow one another as follows:

	Action name	Action type	Dedicated Personnel
Step 1	Rinsing	Simple MD rinsing	One person
Step 2	Washing	Hand washing of MDs such as microsurgical instruments and fragile or non-submergible MDs	One person
		Washing by washer (washing machines)	
Step 3	Packaging	Sorting of equipment according to use and placing in metal boxes according to a use specific to each type of intervention Arrangement in autoclave	Several people

Table 3.1. *Sterilization steps*

The "autoclave" is the machine that sterilizes the MD. In this process, there are several types of flow: the main flow of equipment, the flow of single-use equipment and the flow of loan equipment. Reusable equipment circulates through all steps of the process and constitutes the main flow. Single-use equipment directly enters the packaging step and is discarded after use. As a precautionary principle, the equipment on loan from outside the facility is washed, packaged and sterilized before use. After use, it is washed, packaged and sterilized before being returned to suppliers.

3.3.2. Simulation model

From this diagram, we built a model, based on the ProModel simulation software, with the objective of improving the performance of the sterilization department. We have made several simplifying assumptions: among others, we have neglected urgent containers which must be sterilized as a priority in order to be available as soon as possible in the operating rooms, because their number is not really important. We have counted on average 8 autoclave cycles per month, treating urgent containers compared to 338 autoclave cycles per month (2.37%). We also neglected the flow of single-use and loan equipment because we did not have enough data. Finally, we have simplified the management of scrubbers and autoclaves. The model developed is illustrated in Figure 3.3.

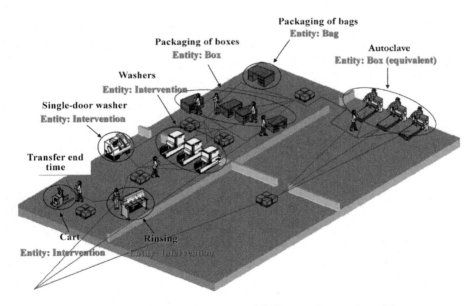

Figure 3.3. *Sterilization process model. For a color version of the figure, please see www.iste.co.uk/sarazin/health.zip*

Upstream of the washers, the entity that moves in the system is called an "action", and after the washers, each action is divided into boxes and bags, the quantity of which depends on the action. At the entrance of the system, we chose to reason in terms of action (rather than in terms of boxes and bags) to be sure that all boxes and bags related to the same action are washed in the same washer. It is indeed this rule that is applied at CHPSM: we, for the most part, try to put all the MDs relating to the same action in a single washer.

After use, the MDs of an action are soaked in trays on a cart. Someone from the OR moves this cart down to the sterilization department. Note that this person is also responsible for cleaning the operating rooms and changing rooms. In our model, this is represented by an entity called "action", entering the place called the "cart". At this point, the MDs wait in storage until the end of the pre-disinfection step (at least 15 minutes) and then wait until the person in charge of rinsing is available. After rinsing, the "actions" pass into the washing stock before being loaded into a washer. If a washer is already partially filled, the action will complete its loading until the washer is filled (its capacity is reached), and at this moment, this washer starts working. Otherwise, a washer is randomly selected to load the following actions. Note that in our model, the capacity of double door washers was set at four actions (this corresponds to the average observed over the period studied). The washer starts working when four actions have been loaded. In reality, washer management is more complex, since not all "actions" occupy the same volume and thus the load is not always four actions. In addition, it sometimes happens that it is not possible to load all the equipment (and containers) used during an operation into a single washer.

After the washing stage, an action is divided into boxes and bags, the quantity of which depends on the action in question. The boxes are placed in the box packaging storage and the bags are placed in the bag packaging storage. There are four box packaging stations; each station can be occupied by one person. One person is in charge of bag packaging and sterilizing them in autoclaves. There are two types of entities that are supplied to autoclaves: boxes and bags. To calculate the capacity of autoclaves, an equivalence is established between these two types of entities. In other words, it was determined how many bags are equivalent to one box. This aims to load a single entity called the "equivalent box" into the autoclaves by studying the actual autoclave load over the period considered. We find that one box is equivalent to six bags, and the bags are then grouped in batches of six. The boxes and batches of bags are then placed in the product storage to be sterilized and are considered as "equivalent boxes", which are loaded into the autoclave, until the nominal load is obtained, to be sterilized. At the exit of the autoclave, the sterilized boxes and bags are grouped in a final stock and are removed from the model.

Note that in this study, we chose to simplify the filling strategy for scrubbers and autoclaves to focus on other areas for improvement. However, it is certain that the optimized management of these resources should make it possible to further reduce the MDs' time spent in the sterilization department. We will address this point in section 3.3.

At the model input, we used real data, from documents called "link cards", which are used to transmit information from the operating room to the sterilization department. The other data required were estimated from the information provided by the CHPSM computer system and are presented in Table 3.2. We also take into account the personnel planning (see (Ngo Cong, 2009) for details). Note that the sterilization department is open from 6:30 to 22:00. Most people work in 8 hour slots: 6.30 to 14:30, 12:00 to 20:00, or 14:00 to 22:00. It should be noted that from 6:30 to 9:00, all staff present pack containers and bags washed the day before, as well as single-use equipment. The last autoclave is started at 20:00.

	Capacity	Average duration
Rinsing	1 action	8 min/action
Single door washer	3 actions	60 min/cycle
Double door washer	4 actions	60 min/cycle
Packaging of boxes	1 box	Normal distribution: Average: 21.2 min/box Standard deviation: 3.16
Packaging of bags	1 bag	1 min/bag
Autoclave	10 equivalent boxes	105 n/cycle

Table 3.2. *Data required in the model*

3.3.3. *Results*

When we study the simulation results obtained with this model (after validation), we observe that 11.9% of MDs aren't sterilized until the day after their use. These MDs are distributed as follows: 2.71% spend the night in washing storage, 4.7% in packaging storage and 4.4% in front of the autoclave. This may explain, in part, the fairly high value of retention time. Indeed, the residence time is about 8 hours. If we add up the minimum soaking time and the estimated duration of the other stages (rinsing, washing, packaging and sterilization), we obtain about 3 h 30 min. The rest (about 4 h 30 min) is the time spent in intermediate storage, plus the time at night for some MDs. It should be noted that, in the current situation, the activity stops at 20:00 at the rinsing station and the washing station, and at 22:00 at the packaging station and the sterilization station, which may also partly explain why a significant part of the MDs that arrive in the sterilization department at the end of the afternoon still remain in storage.

Following these results, we proposed to test different scenarios by modifying the system input parameters, the personnel planning, and we compared the performance of the system obtained for each scenario. We have thus been able to show that performance can be significantly improved by streamlining the arrival flow of MDs (the implementation of a periodic transfer of MDs from the operating room to the sterilization department significantly reduces the residence time and the duration of the pre-disinfection of MDs), or by modifying the personnel planning and the opening hours of the department (the implementation of a new personnel planning obtained, with a constant workload, by shifting personnel to the evening, by increasing personnel at the time when the activity is at its peak, and by decreasing it at the time when it is at its lowest, made it possible to reduce the retention time of the MDs and to have practically no more non-sterilized MDs at the end of the day). The results obtained in this study prompted those in charge of the sterilization department to re-examine the organization of their department and gave them arguments to justify the changes to CHPSM management.

After this study specific to a particular sterilization department, we were interested in comparing the performance of certain management practices used in central sterilizations in the Rhône-Alpes region through a survey and the development of a generic simulation model.

3.3.3.1. *Comparison of several sterilization departments*

Although a certain number of ministerial recommendations concerning hygiene allow an almost identical sterilization process to be observed from one establishment to another, on the other hand, as regards the organization of production, each sterilization department has total freedom, and practices are diverse. This observation prompted us to set up the 2E2S project (Electronic survey of sterilization departments) to draw up an inventory of the organizational sterilization practices of establishments in the Rhône-Alpes region and to identify isolated practices. These isolated practices can be ineffective practices inherited from the past but also innovative practices, "experiments" initiated by sterilization teams in search of a better management of their department.

The survey conducted as part of this project led to the drafting of a report (Reymondon *et al.* 2008a) *which presents and analyzes the data collected on the organization of sterilization departments, and a publication* (Reymondon *et al.* 2008b), *which are summarized here.*

This work also resulted in the development of a generic simulation model for a sterilization department and a comparison of the performance of different facilities that responded to the survey, details of which can be found in (Ngo Cong 2009) *and* (Di Mascolo and Gouin 2013).

3.3.3.2. *Survey on the organization of sterilization departments*

The survey was based on an online questionnaire. This questionnaire was sent to 75 establishments belonging to the categories "public establishments", "private establishments" and "private establishments participating in the public hospital service". Of the 39 respondents, 23 had central sterilization and 16 had non-centralized sterilization. Here, we are only interested in central sterilization departments. We finally selected 14 of these departments, corresponding to those that answered all the questions. They are named H1 through H14 in the following paragraph. We have classified them by volume of activity (in autoclave DIN baskets per day − a DIN basket is a standard washing volume of $480 \times 260 \times 50$ mm, or a sterilization volume of $600 \times 300 \times 300$ mm) decreasing before naming them, so H1 is the establishment with the highest volume of activity and H14 is the one with the lowest volume of activity.

Centralized department data from the survey are responses to closed-ended yes/no questions, quantitative questions, multiple choice questions or comments. The electronic survey was structured into six parts: preliminary information, general information, schedules, equipment resources, human resources and sterilization processes. The first part, "preliminary information", gathered general information on the establishment and on the department. The general activity was then requested in the second part entitled "Generalities". The third part gathered information about the operating room and sterilization department schedules. In the "equipment resources" section, information was obtained on the equipment for all steps of the sterilization process, for example, rinsing, washing, packaging and autoclave sterilization. The fifth part concerned "human resources" (number, guards, on-call duties, etc.). Finally, the last part asked questions about the organization of the sterilization process. We can see some raw survey results in Table 3.3.

	H1	H2	H3	H4	H5	H6	H7	H8	H9	H10	H11	H12	H13	H14
Volume of activity (in autoclave DIN baskets per day)	235.9	129.7	123.1	112.5	101.3	93.6	89.8	74.3	73.6	59.2	25.4	24.0	22.2	16.2
Average number of autoclave loads per year	4600	4600	4000	3987	4388	4295	1946	2759	4100	2200	1100	892	960	700
Sum of washer-disinfector capacities: total number of baskets (DIN)	50	24	24	28	30	20	21	15	26	14	8	6	15	16
Sum of autoclave capacities: total number of baskets (DIN)	40	22	24	22	18	17	24	14	14	14	12	14	12	12
Personnel: total number of full time equivalents (FTEs)	25.0	9.0	14.0	13.0	13.4	13.0	6.3	9.8	12.6	8.0	5.5	4.5	3.5	3.0
Open hours (hours)	14.0	13.0	15.5	13.5	13.0	13.5	11.5	15.0	13.0	14.0	9.6	12.5	9.5	8.0

Table 3.3. *Examples of raw survey results (Ngo Cong 2009)*

The multi-view analysis of the survey results (that is, a view of the sterilization activity, a view of the MD production system, a view of the sterilization process, etc.) showed that centralized sterilization departments often have similar practices and organizations. This led us to develop a basic generic model, presented in the following section, from which we can evaluate performance through simulation. We also found some isolated practices. The question that then arose was whether these were innovative practices that provided real added value in terms of performance or whether they were local practices inherited without any real interest. Finally, the survey highlighted that, perhaps, new and innovative practices could be considered to improve the performance and duration of the production cycle. To take just one example, most of the washers used (81% of the survey responses) allow washing and rinsing. However, 73% of the departments said they did a manual rinse. Therefore, in 57% of cases, a double rinse is performed. It is therefore possible to consider eliminating the rinsing stage, as the instruments can then be placed in the scrubbers as soon as they enter the sterilization department, if the pre-disinfection time can be controlled. We explore this possibility in section 3.3.

3.3.3.3. *Generic model for the performance evaluation of sterilization departments*

In our specific simulation, presented in section 3.1, we made assumptions about the structure of the model and the data used. In the generic simulation (which we carried out with the ARENA software), we used the same assumptions on the structure, but additional simplifications were necessary to be able to use the survey data. Our simplifications concerned the input data (arrival of medical devices at the sterilization department), the personnel staffing and the loading rule for the washers. The other data (capacity of each position, working time on each position in the system) were directly obtained from the responses to the survey questions for each department.

In terms of system input data, we only had the number of sterilized DIN baskets per day, and we had no indication of the distribution of these arrivals during the day. We used the same entry profile for all departments to be compared. To establish this profile, we used the data we had collected at CHPSM (number of interventions received each hour, volume of these interventions in DIN baskets and constitution of these interventions in boxes and bags). The profile obtained for these input data was used for the different departments compared, taking into account the volume of activity of these departments. We also compared the results obtained using this profile with the results obtained using extreme profiles, for example, a uniform distribution of arrivals during the day, very important peaks at certain times of the day, etc.

As far as staff planning was concerned, we only had the total number of people in the department and we did not know the distribution of the workload of these people during the day. We used a staff schedule with constant work potential throughout the day. Based on the number of people working in the sterilization department and using a working time of 8 hours for each person, we calculated the total number of hours worked each day. The number of people working each hour was then calculated by dividing the total number of hours worked each day by the opening hours. This made it possible to identify the number of people present in the department for each hour, then to assign two people to washing and rinsing tasks, one person to driving the autoclaves and the remaining people to packaging.

Finally, with regard to washer loading, the investigation showed us that the loading rules used are not the same for all sterilization departments. Some establishments use rules based on a filling threshold: load maximization (threshold to be tested: percentage of filling), others use rules based on a waiting time threshold: a minimization of upstream stocks (threshold to be tested: waiting time between arrival and washing). We used these two rules in our simulations, according to the response given to the survey, by considering a washing capacity expressed in the number of DIN baskets.

After its validation on the results of the CHPSM, we used this generic model to evaluate the performance of nine sterilization departments that responded to our questionnaire in the list of the 14 previously selected. These establishments are H1, H2, H3, H4, H5, H6, H8, H9 and H10. The excluded establishments are those that have very few people working in the sterilization department and therefore certainly have staff working several shifts in the same working day, which we cannot take into account in our model.

3.3.3.4. *Performance comparison of several sterilization departments*

Figure 3.4 illustrates the results obtained when observing the retention times in the different departments studied, distinguishing the total retention time from the time spent in storage. It also shows the proportion of the time spent in storage in relation to retention time.

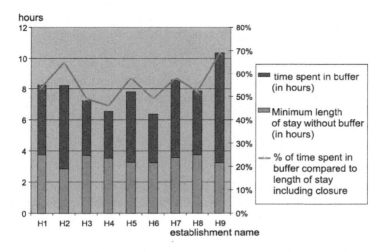

Figure 3.4. *Results obtained for residence time. For a color version of the figure, please see www.iste.co.uk/sarazin/health.zip*

Figure 3.5 shows, for each facility, the number of boxes processed in a day and those not sterilized at the end of the day. It also shows the proportion of containers remaining unsterilized at the end of the day, relative to the total number of containers processed.

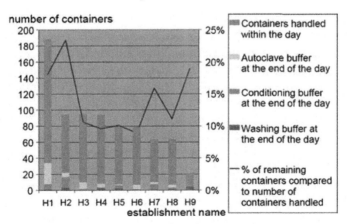

Figure 3.5. *Number of non-sterilized containers at the end of the day. For a color version of the figure, please see www.iste.co.uk/sarazin/health.zip*

These two figures illustrate the fact that performance considerably varies from one establishment to another and that it is not necessarily the largest departments that have the best performance.

More generally, the results obtained by the generic model have shown the interest of such a tool to obtain performance indicators that are difficult to estimate by sterilization department managers (Di Mascolo and Gouin 2013).

3.3.3.5. *Optimization of washing resources*

We now consider the problem of loading washers in the washing stage of a sterilization department, which had been very simplified in the models previously presented and which, according to the survey mentioned previously, constitutes a bottleneck.

In a hospital, the number of MD referrals is generally high (several hundred). All the MDs used for an intervention form the MD set of that intervention. Since each intervention may require different numbers and types of MDs and the MD sets may be of different sizes. In addition, for various reasons (duration of interventions, general organization of the sterilization department, etc.), the MD sets are available for washing at different times of the day.

For washer loading, there are two main categories of decisions to be made: how to form batches with different sets of MDs and when to start a wash cycle. For organizational reasons, it is not permitted to cut MD sets to assign them to different wash cycles. Otherwise, the downstream washing steps may be complicated because of the large number of MDs in the sets and their similarities.

One strategy, often used in sterilization departments, is maximizing washer load. For this, a filling threshold is determined. Then, based on this threshold, a batch is created with sets of MDs as they arrive. When an MD set enters the wash stock, if there is a washer available, the baskets of that set are placed in the washer. Each time a new set of MDs arrives, the baskets are placed in the washer. As soon as the filling threshold is reached, a washing cycle is started. This threshold is around 80–90% of the washer capacity. Of course, if there are still MDs available, and the filling threshold can be exceeded while respecting the capacity of the washer. If there is no washer available when the MDs arrive, they are placed in a holding batch.

Considering that there is no manual rinsing in the system, the application of this strategy can lead to a long waiting time for the MD in the wash stock. Of course, the longer the waiting time, the longer the pre-disinfection time increases. We therefore need other strategies, more sensitive to the duration of MD pre-disinfection. Different performance criteria can be considered for washing operations, such as minimizing the total wash time, minimizing the number of wash cycles initiated, minimizing the average waiting time, etc.

In Ozturk's thesis (2012), we proposed several efficient strategies to load the washers and thus improve the MD flow. We started by considering the simplified case for which we consider that we know in advance all MD arrivals and for which we proposed offline methods to minimize the total washing time (Ozturk *et al.* 2012a), or to minimize the average waiting time for washers (Ozturk *et al.* 2010), or to minimize the number of wash cycles (Ozturk *et al.* 2011a). We then took into account the uncertainties of MD arrival times and developed semi-online or online methods to minimize MD pre-disinfection time (Ozturk *et al.* 2012b).

Finally, we integrated the developed algorithms into the generic model presented above to test their impact on the performance of the sterilization department as a whole (Ozturk *et al.* 2011b).

In particular, we were able to show that it was possible to guarantee a reasonable pre-disinfection time, thus making it possible to eliminate the manual rinsing stage and to transfer the resources released during rinsing to the conditioning station, which turned out to be the new bottleneck. The numerical results showed that in the absence of manual rinsing and in the presence of an additional conditioning station, it was possible, with the developed algorithms, to increase the number of sterilized MDs per day, while guaranteeing good pre-disinfection times.

3.3.3.6. *Taking risks into account – simulation in degraded mode*

In all the work we have presented, we have seen how to improve the performance of a sterilization department using performance evaluation. In performance assessment models, however, the occurrence of risks is generally only incorporated to a limited extent. Most of the time, we simply put random durations to take into account the effect of possible hazards and uncertainties. However, the occurrence of a failure can lead to a change in the structure of the system (for example, when a machine fails, a degraded mode can be put in place, using other resources while waiting for the repair or replacement of the failed machine). In this case, the performance evaluation of the degraded solution will be done by modifying the original model, for each of the possible risks, and by evaluating the performance of the possible new solutions. This approach has the disadvantage of being cumbersome to implement, especially when a production system presents a large number of risks. It does not easily represent all system failures, does not take into account the propagation of failures and does not link each failure to the performance indicators it influences.

In addition, several studies have been conducted in healthcare institutions in order to identify possible risks in these organizations. We can cite, for example, the work of Bernard and Lacroix (2001), which are part of a project to restructure the *CHU* de *Haut-Lévêque* (Haut-Lévêque University Hospital) in Bordeaux. The aim of this work was to improve the quality of MD processing and was based on a risk analysis in the central sterilization unit. We can also quote Talon's work (2011) that was conducted in the *Hôpital Bichat* (Bichat Hospital) in Paris. Using a risk analysis, they made it possible to develop a new traceability system for MDs to reduce the risks identified using the analysis performed.

However, the risk analysis work carried out only takes into account the deterioration in department performance when a risk occurs, in a limited way. This deterioration is assessed on the basis of qualitative ratings that lack precision. As for the performance evaluation work, we have seen that it either considers that no failure occurs or that there is an occurrence of risks, but without taking into account certain important aspects, such as the propagation of failures in the system, or the dynamic change of the flow configuration in the event of a major failure.

Negrichi's thesis (2015) aimed to create a tool to automatically evaluate the performance of a production system in degraded mode (that is, taking into account the occurrence of risks), based on a model that organizes information, describing the system's behavior and architecture, as well as its failure modes. The approach is illustrated in Figure 3.6. This tool was applied to the case of the sterilization department.

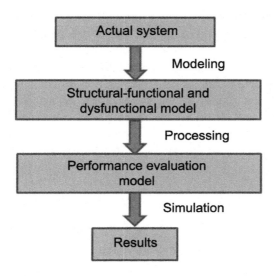

Figure 3.6. *Proposed approach for the analysis of production systems (Negirichi* et al. *2013)*

Our main idea was to create a basic model (structural-functional and dysfunctional model) which makes it possible to group information on the behavior of the system (system functions and resources) in its various possible states (normal mode or degraded mode following the occurrence of a risk) (Negrichi *et al.* 2017).

We used the FIS (risk analysis method) approach, which is a model-driven risk analysis method (Flaus 2008), which allowed us to develop a risk model for sterilization departments, based on a field study and a literature review (Negrichi *et al.* 2012). This SIF model was then automatically transformed into a performance evaluation model (using a translation algorithm) (Negrichi *et al.* 2014). We chose to use (and adapt) the *Réseaux de petri à haut niveau* (the high-level petri networks) (to which we added the necessary functionalities), in order to obtain a model capable of representing the possible evolutions of the system (in normal and degraded mode). A simulator was then developed from this *Réseaux de petri* in order to calculate the relevant performance indicators (Di Mascolo *et al.* 2017).

The entire proposed approach was applied to the sterilization department of a hospital in Grenoble. We carried out a detailed risk analysis, which revealed 177 dangerous phenomena, most of which concerned the MD packaging and sterilization stages, and 536 failure modes, mainly concerning the rinsing and packaging steps. We then applied our simulator to a somewhat simplified model of the sterilization department. We were thus able to observe the effect on performance of the occurrence of breakdowns on scrubbers, or the spread of contamination of certain MDs, or taking a degraded mode of behavior into account, in addition to normal behavior.

3.4. Our outlook

Several research axes emerge from the work that has been presented here:

– the tool resulting from Negrichi's thesis is generic and was developed for experimental purposes. We would now like to improve its usability and make it more field oriented;

– for the moment, this tool allows us to only study how failures spread. We think it would be interesting to include the possibility of detecting failures, which would stop the spread;

– finally, it seems interesting to us to integrate optimization approaches into this tool to optimize performance in degraded mode.

3.5. References

Bernard, V. and Lacroix, P. (2001). Restructuration d'un service de stérilisation dans le cadre d'une démarche qualité. *ITBM-RBM,* 22(2), 116–124.

Di Mascolo, M. and Gouin, A. (2013). A generic model for the performance evaluation of sterilization services in health establishments. *Health Care Management Science*, DOI 10.1007/s10729-012-9210-2, 16(1), 45–61.

Di Mascolo, M., Flaus, J.M., and Negrichi, K. (2017). A simulation tool to assess the performance of production systems in gegraded mode. *11th Conference on Stochastic Models of Manufacturing and Service Operations*.

Flaus, J.M. (2008). A model-based approach for systematic risk analysis. *Journal of Risk and Reliability*, 222, 79–93.

McNally, O., Thompson, I.M., McIlevenny, G., Smyth, E.T.M., McBrice, N., and MacAuley D. (2001). Sterilization and disinfection in general practice within university health services. *Journal of Hospital Infection*, 49, 210–214.

Ministère de l'emploi et de la solidarité (2001). Bonnes pratiques de pharmacie hospitalière. Technical report [Internet]. Available at: http://www. ladocumentationfrancaise.fr/rapports-publics/014000475/index.shtml.

Negrichi, K. (2015). Approche intégrée pour l'analyse de risques et l'évaluation de performances : application aux services de stérilisation hospitalière. PhD thesis, University Grenoble Alpes.

Negrichi, K., Di Mascolo, M., and Flaus, J.M. (2012). Risk analysis in sterilization services: a first step towards a generic model of risk. *6ème conférence francophone en gestion et ingénierie des systèmes hospitaliers*, Quebec, August 30–September 1.

Negrichi, K., Di Mascolo, M., and Flaus, J.M. (2014). Conversion of a risk model into a petri net model for simulation and analysis. *European Safety and Reliability Conference*, ESREL.

Ngo Cong, K. (2009). Etude et amélioration de l'organisation de la production de dispositifs médicaux stériles. PhD thesis, Joseph Fourier University.

Ngo Cong, K., Gouin, A., Di Mascolo, M., and Schwob L. (2007). Etude d'un service de stérilisation de dispositifs médicaux. *Gestions Hospitalières*, 465, 278–285.

Ozturk, O. (2012). Optimisation du chargement des laveurs dans un service de stérilisation hospitalière : ordonnancement, simulation, couplage. PhD thesis, University of Grenoble Alpes.

Ozturk, O., Di Mascolo, M., Espinouse, M.L., and Gouin, A. (2010). Minimizing the sum of job completion times for washing operations in hospital sterilization services. *8ème Conférence internationale de modélisation et simulation*, Hammamet, Tunisia, May 10–12.

Ozturk, O., Espinouse, M.L., Di Mascolo, M., and Gouin A. (2012a). Makespan minimisation on parallel batch processing machines with non-identical job sizes and release dates. *International Journal of Production Research*, 50(20), DOI: 10.1080/00207543.2011.641358, 6022–6035.

Ozturk, O., Di Mascolo, M., Espinouse, M.L., and Gouin A. (2012b). A semi-online algorithm for optimizing the pre-disinfection duration of medical devices in a hospital sterilization service. *9ème Conférence internationale de modélisation et simulation*, Bordeaux, 8–10 June.

Ozturk, O., Sebö, A., Espinouse, M.L., and Di Mascolo, M. (2011a). An optimal bin packing algorithm to minimize the number of washing cycles in a hospital sterilization service. *International Conference on Operational Research Applied in Health Services, ORAHS 2011*, Cardiff, United Kingdom.

Ozturk, O., Di Mascolo, M., Espinouse, M.L., and Gouin, A. (2011b). Optimisation du chargement des laveurs dans un service de stérilisation. *4ème Journée doctorales/journées Nationales MACS*, Marseille.

Reymondon, F., Pellet, B., Calleja, G., Marcon, E., Di Mascolo, M., and Gouin, A. (2008a). Rapport d'enquête du projet 2E2S - Etude organisationnelle des services de stérilisation Rhônalpins. Report, France.

Reymondon, F., Di Mascolo, M., Gouin, A., Pellet, B., and Marcon, E. (2008b). Etat des lieux des pratiques de stérilisation hospitalière en Rhône-Alpes. *4ème Conférence francophone en gestion et ingénierie des systèmes hospitaliers*, Lausanne, Switzerland.

Rutala, W.A. and Weber, D.J. (2004). Disinfection and sterilization in healthcare facilities: what clinicians need to know. *Clinical infectious diseases*, 39(5), 702–709.

Smyth, E.T.M., McIlvenny, G., Thompson, I.M., Adam, R.J., McBrice, L., Young, B., Mitchell, E., and MacAuley D. (1999). Sterilization and disinfection in general practice in Northern Ireland. *Journal of Hospital Infection*, 43, 155–161.

Talon, D. (2011). Gestion des risques dans une stérilisation centrale d'un établissement hospitalier : apport de la traçabilité à l'instrument. PhD thesis, Ecole Centrale Paris, France.

4

Prediction of Hospital Flows Based on Influenza Epidemics and Meteorological Factors

4.1. Introduction

In France, hospitalization flows partly depend on admissions made by emergency services. Their prediction is done in a contextual way each day based on empirical data (Carli 2013). However, these fluctuations may be related to weather variations. Indeed, the influence of the weather on health has long since been recognized; in the 5th Century BC, Hippocrates had already written: "Do not ignore what affects the weather has, because everything that concerns it is closely related to medicine." (BDSP, 2012). This relation is much more obvious when it comes to viral infections and influenza in particular (Altizer 2013; Fleming 2007).

DEFINITION.– Seasonal or common influenza is an easily transmissible viral infection that progresses rapidly into an epidemic and especially in winter. It is characterized by the occurrence of a fever above 39°C, of sudden onset, accompanied by myalgias and respiratory signs (cough and bronchial congestion). The virus causing influenza contains deoxyribonucleic acid, and there are four types of this virus, namely A, B, C and D. These types have subtypes determined by surface proteins present on the surface of the peripheral envelope of the virus: hemagglutinin (H) and neuraminidase (N).

Chapter written by Radia SPIGA, Mireille BATTON-HUBERT and Marianne SARAZIN.

For example, type A can be H1N1 or H5N1. These subtypes will be at the origin of the more or less virulent characteristics of the virus.

Although it is benign and generally progresses towards spontaneous recovery, influenza nevertheless remains a major public health problem associated with significant morbidity and mortality among at-risk individuals as well as an increase in hospitalization flows at the time of epidemic peaks, requiring the deployment of material and human resources (Dao and Kamimoto 2010; Mertz 2013; Zhou 2012; Jhung 2014). In France, an influenza epidemic is declared when its incidence exceeds two consecutive weeks as well as the threshold established by the Sentinels network based on the Serfling model (Réseau Sentinelles 2018; Serfling 1963).

Despite clinical knowledge and the existence of a vaccine against influenza, it continues to "emerge" each year with epidemic seasons of varying duration and intensity, making it difficult to anticipate management (Cannell 2008; Bresee 2013). In order to explain the seasonal nature of influenza, several hypotheses have been proposed: seasonal variations in the host's immune state are linked to climatic variations (sunlight and low temperatures) and the transmissibility of the virus depends on temperature and humidity. The effect of sunlight is via UVB rays, which stimulate photosynthesis of vitamin D whose serological concentration was found to be at its highest in August and lowest one month after the winter solstice; vitamin D plays a role in regulating acquired immunity and strengthening innate immunity (Cannell 2006; Beard 2011; Borella 2014). The role of temperature in virus transmission was highlighted in the study by Lowen *et al.* (2007) on guinea pigs, with higher transmission at temperatures below 5°C. In addition, cold has a direct effect on the nasal mucosa, the first physical barrier to infection, by increasing the viscosity of mucus, decreasing mucociliary clearance and promoting the spread of the virus in the respiratory tract, with maximum stability of the virus in the upper respiratory tract under the same conditions (Lowen *et al.* 2008). Finally, a low level of humidity would promote the formation and stability of condensation nuclei (droplet nuclei) containing the virus, with an increase in exposure time and thus virus transmission (Tellier 2009; Noti 2013).

Several modeling studies of influenza epidemics have been carried out taking weather factors into account (Shaman 2009a; Shaman 2009b; Shaman 2011; Shaman 2012; Axelsen 2014; Tamerius 2013). Few have addressed the link between influenza and emergency room admission flows (Foster

2013; Morina 2011; Schanzer 2013; Sahu 2014); among them, the study by Sahu *et al* (2014) estimated the influence of weather variations on emergency admissions by taking the case of influenza epidemics, and the study by Foster *et al.* (2013) demonstrated a relationship between the seasonal nature of influenza epidemics and myocardial infarction.

This work builds on the results of previous studies, using factors identified in the literature to predict influenza cases seen at the hospital. Two approaches were used: first, discriminant factor analysis (DFA), which takes into account meteorological factors and general practitioners' epidemic data, and second, the probability of transition from one epidemic state to another at the hospital level, which is calculated using the Markov chain method.

This study was carried out using data from the Loire administrative government department and assesses the impact of influenza epidemics and meteorology on hospitalizations. It is a first step in helping to predict the activity of hospital emergency services.

4.2. Method

The study was carried out in a circumscribed geographical area and all the information necessary for its completion was, as far as possible, related to this area. Several methods were used to define the impact of each factor studied on the volume of hospitalizations in the main public health facilities in this geographical area.

4.2.1. The data

All data were collected over a period ranging from the first week of 2007 to the eighth week of 2015 and over a circumscribed geographical territory: the south of the Loire administrative governmental department. They come from three sources:

The meteorological data come from the station of Andrézieux managed by Météo France (www.meteofrance.com/accueil). The variables considered were daily mean temperature in °C, relative humidity (RH) in %, and daily sunshine duration. The absolute humidity (AH) was calculated from the temperature and RH, while the sunshine in w/m² was calculated from the sunshine duration.

Influenza data observed in general practitioners were provided by the Sentinels network (Réseau Sentinelles). It is a network made up of 1,300 independent general practitioners and volunteers, spread over the French metropolitan territory, continuously collecting information on eight health indicators derived from their activity, including influenza.

The hospital data come from the PMSI databases (French National Hospital Database) from the hospitals *CHU de Saint-Etienne* (Saint-Etienne University Hospital) and *CH de Firminy* (Firminy Hospital). Two types of data are noted: hospitalizations associated with influenza (IAH) and emergency room entry data.

4.2.2. Data processing

In order to respect the unit of time used by the Sentinels network, meteorological data were aggregated into weekly averages, and hospital data into the number of cases per week.

A hospital epidemic threshold was characterized for each week by taking the quartiles of the influenza-associated hospitalizations (IAH) variable. A categorical variable was thus constituted taking the value 0 or the non-epidemic state for case volumes below the first quartile, the value 1 or the intermediate state (volumes between the first and the third quartile), and the value 2 or the epidemic state (volumes above the third quartile).

4.2.3. Data analysis

First, the data were described by their means, standard deviation and extreme values.

The year 2009 (year of the influenza A/H1N1 pandemic) has been withdrawn from the following analysis steps.

Next, the relationships between IAH and other data, such as meteorological, emergency and Sentinel data, were measured using Pearson's correlation by testing time lags, to take into account the possible time influence of one factor on the other. This analysis was consolidated by an analysis of variance (ANOVA) to compare the data according to each group, the weeks' epidemic state and the age classes.

A principal component analysis (PCA) (Husson XX) allowed the data table (n individuals/p variables) to be described by presenting them in a reduced extent, thus highlighting possible structures within the data. The similarity between individuals was assessed by the distances between them, and the correlations between the different variables were estimated by the linear correlation coefficient. The n individuals correspond to weeks: from the first week of 2007 to the eighth week of 2015, excluding the 2009 pandemic year, that is, 372 weeks.

A hierarchical upward classification of the Ward method consolidated by a classification by moving averages (k-means) made it possible to group the individuals having the same profile and to separate the various profiles.

The predictive step

First, a DFA was performed (McLachlan 2004). It is a method that describes, explains and predicts the membership of predefined groups (classes) of a set of observations (in our case, the weeks of the year) from a series of predictive variables (weather variables and Sentinel variables).

Finally, to estimate the probability of transition of future weeks from one epidemic state to another, a Markov chain model was used (Davies 1975). A Markov chain is a stochastic process with discrete time and discrete state space, having Markov's properties: the ability to predict the future knowing the present.

Data processing and analysis were carried out using the R 3.1.2 software.

4.3. Results

The results of the study consider several aspects: on the one hand, a description of the behavior of each factor in the area studied, then the concomitant variations between these factors and, on the other hand, the influence of each factor on hospital flows when they are analyzed together.

4.3.1. *Description of the data*

Between 2007 and 2015, there were 11,389 IAHs (with a minimum of three IAH/week and a maximum of 104 IAH/week); the number of hospitalizations varies according to years and seasons (Figure 4.1) with a weekly average of 31 IAHs in winter and 19 IAHs during the rest of the year, and according to age classes with a significantly higher number for those over 65 (Figure 4.1). During the same study period, the mean values (and standard deviations) of temperature, AH and sunshine in the Loire administrative government department were 11.33°C (6.76), 7.68 g/m3 (0.65) and 185.80 w/m² (128.76), respectively.

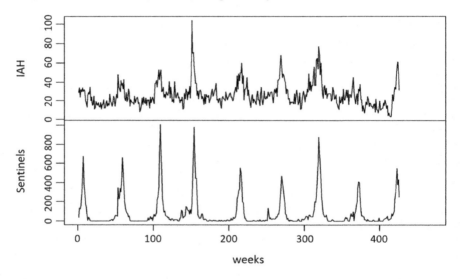

Figure 4.1. *Top: influenza and influenza syndromes observed in hospitals from week 1 of 2007 to week 8 of 2015. Bottom: influenza and influenza syndromes observed in general practitioners from week 1 of 2007 to week 8 of 2015*

4.3.2. *Correlations between variables*

A negative correlation exists between IAH and meteorological factors (p < 0.001); it is more significant when considering a time lag of -13 days for temperature, -12 days for AH and -32 days for sunshine (Table 4.1).

A good correlation is observed with the ED hospitalization variable (0.53; p value < 0.001), as well as with the Sentinels network data (0.68;

p value < 0.001). The latter increases when the Sentinel data are shifted by −1 week (0.70; p value < 0.001).

	Sentinels	Sentinels (−1 week)	Single passages	Urgent intake	Temperature (−13d)	Humidity abs (−12d)	UV light (−32d)
IAH	0.68	0.70	−0.01	0.53	−0.63	−0.61	−0.55
P value	< 0.001	< 0.001	0.9	< 0.001	< 0.001	< 0.001	< 0.001

Table 4.1. *Pearson correlation coefficients between the IAH variable and the other variables, taking into account time lags*

4.3.3. ANOVA

Among IAH cases, the over-65 age group is significantly larger than the other age groups (Figure 4.2), in contrast to urban influenza cases where the under-65 age group is significantly predominant ($p < 0.001$). In addition, a difference was also found for each meteorological data according to the existence or not of an epidemic state in hospitals: the values measured for each meteorological variable are significantly lower for the weeks considered as epidemic ($p < 0.001$).

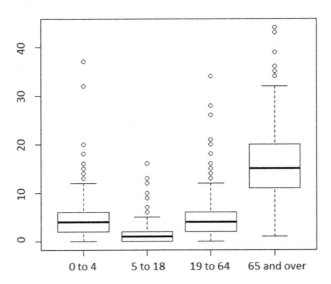

Figure 4.2. *ANOVA – IAH by age group*

4.3.4. *Principal component analysis*

The first two PCA factorial axes explain 69.5% of the information and 55.03% by axis 1 with a greater contribution of weather variables: temperature and AH. The projection of the variables on the plan of the first two PCA factorial axes shows an anti-correlation between meteorological variables and clinical variables (IAH, Sentinel data and emergency data). There is no correlation between single emergency passages and other variables (Figure 4.3).

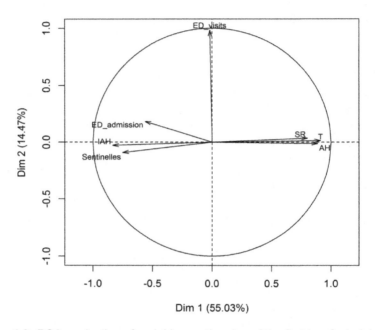

Figure 4.3. *PCA: projection of variables on the plan of the first two factorial axes. IAH: influenza-associated hospitalizations; ED-admission: admission from emergency services; ED-visits: consultations in emergency departments without hospitalization*

4.3.5. *Classification*

The hierarchical ascending classification and the k-mean method classified individuals into three cluster groups (Figure 4.4). These groups differ mainly in temperature factor: the first group consists mainly of epidemic weeks with an average temperature (and standard deviation) of 2.62°C (3.66), the second group corresponds to weeks of intermediate epidemic status with an average temperature of 7.62°C (3.57) and the third group consists mainly of non-epidemic weeks with an average temperature of 17.47°C (2.88).

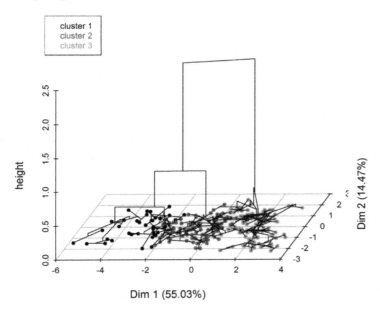

Figure 4.4. *Hierarchical ward classification consolidated by K-means classification. For a color version of the figure, please see www.iste.co.uk/sarazin/health.zip*

4.3.6. Discriminant factor analysis

The first part is descriptive, allowing us to visualize the individuals in each group as they were discriminated against by DFA functions. Axis 1 explains 95.97% of the information and discriminates well between the group without epidemic and the group with epidemic, while axis 2 does not separate the three groups. The variable with the highest coefficient is the influenza incidence represented by the Sentinel data.

The second part is predictive: the instructive sample consists of week 1 of 2007 to week 8 of 2015. Epidemic weeks are correctly predicted. However, six prediction errors were observed (11% errors): 5 weeks without epidemic were wrongly classified as intermediate epidemic weeks, and 1 intermediate epidemic week was classified as an epidemic week (Figure 4.5).

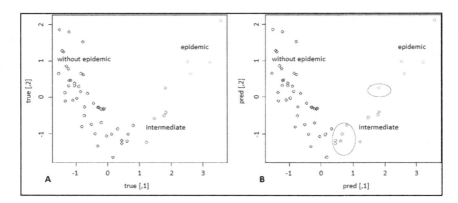

Figure 4.5. *Predictive DFA: prediction sample weeks projected to DFA plan.
A: true epidemic states, B: predicted states, prediction errors circled. For
a color version of the figure, please see www.iste.co.uk/sarazin/health.zip*

4.3.7. Markov chain prediction

When it is winter, the three states are possible (epidemic, intermediate or
non-epidemic) with lower probabilities to pass through the week without
epidemic or to remain in it.

The probabilities of transition from one epidemic state to another are
different depending on weather conditions. When the three weather variables
are very low (32 w/m², 2°C and 4 g/m³ for sunshine, temperature and AH,
respectively), the probability of going directly from one week without
epidemic to one week with epidemic and of remaining so is 1. In a situation
where the temperature, AH and sunshine are higher than the highest
threshold (74 w/m², 6°C and 5.8 g/m3 for sunshine, temperature and AH,
respectively), the probability of going into an epidemic week becomes zero.

4.4. Discussion

This study, carried out in France under continental climatic conditions,
presents methods for characterizing the links between meteorological
factors, the incidence of influenza treated outside hospitals and its
repercussions on hospitalizations for influenza and its consequences. This
approach, which has not been identified to date in the literature, has
confirmed the results already published (Lowen 2007, Shaman 2009a; van

Noort 2012). This work also proposes a predictive method for anticipating influenza management in hospitals.

The strong correlation between influenza cases seen by general practitioners and hospital admission flows for influenza confirms the importance of prediction in the management of influenza epidemics. Hospitals are increasingly congested by an influx of patients, mainly in winter, and their management requires predictive tools in order to adjust the necessary means of management (Carli; Schanzer 2013). While the Sentinels network in France aims to characterize an epidemic state in the general population, this study proposes a method to characterize an epidemic state in a hospital environment.

As has been shown, the effects of meteorological factors on an influenza epidemic in both extra-hospital and hospital settings are with some clinical and scientifically consistent lags. A period of 32 days between sunshine and the number of IAH can be explained by the time required for photosynthesis and the passage of vitamin D into the bloodstream (Cannell 2006). Concerning temperature and AH, they act on the viability and transmissibility of the virus. Taking into account an incubation time of 48 hours (*WHO | Influenza (Seasonal)*), and a latency time before going for a consultation, the intervals before hospitalization observed for these two factors seem high (13 and 12 days). However, the preponderance of patients over the age of 65 may explain these delays. Indeed, the advanced age may be at the origin of complicated clinical manifestations occurring after an episode of acute and late influenza requiring secondary hospitalization (Dao 2010; Jhung 2014; Olson 2007). This hypothesis is supported by a better correlation found when the external consultation data from the Sentinels network precede the IAH data by one week (Pearson correlation = 0.70; Table 4.1).

In order to predict IAHs, a first approach was to use DFA respecting the time intervals previously identified, the aim being to predict the epidemic state of future weeks knowing the values of the meteorological and Sentinel explanatory variables. The prediction was made for the year 2014/2015, with an error rate of 11%, confusion being more significant when it comes to classifying weeks without epidemics, while all the weeks of epidemics were predicted correctly. The second envisaged approach is Markov chain prediction, with an estimate of the probabilities of transition from one epidemic state to another in the winter period based on the states of previous weeks. Better accuracy is obtained when weather conditions are known.

In the hypothesis of hospital use for organizational purposes, these approaches are fully compatible with the way the different data producers (Météo France and the Sentinels network) operate, each variable being provided within deadlines allowing the use of these methods. Data from the Sentinels network staggered by one week, which represents the best predictor with the highest coefficient, allowing the design of computerized tools for hospitals to be considered. However, this approach is dependent on the production of hospital data, which is still late compared to the forecast deadline, varying from 30 to 60 days. The implementation of the *Facturation individuelle des établissement de santé*, FIDES (French individual healthcare invoicing plan), which will soon be adapted to hospital stays, will probably shorten these delays. This is known as the Fides project.

However, this study has some limitations. Only seven years were used, corresponding to seven epidemic events; the integration of additional years would reinforce the reliability of the model and a better perspective on result analysis. Furthermore, no correlation has been demonstrated between IAHs and data from simple emergency department visits. The integration of all diagnoses may explain this result. Unfortunately, as the reasons for consultation were not given, it was not possible to discriminate cases for single passages. The type of virus predominating each year may have consequences in terms of clinical virulence, which would be interesting to take into account. Other factors that may also be linked to epidemics have not been tested, such as the behavior of the population: mode of transport, location and type of work, hand washing or vaccination (Willem 2012; Garza 2013; Beest 2013). These hard-to-gather elements are involved in the spread of an epidemic and could strengthen the predictive capacity of the model considered at the population level and not at the individual level.

4.5. Conclusion

The results confirm the links between IAH, weather variations and general practitioner activity. The study shows that the latter is the best predictor of hospital activity related to influenza. This study is a conclusive first step in considering further work to predict the intensity and duration of epidemics using a quantitative approach. The final step would be to be able to propose a probabilistic model adapted to the hospital and used as a tool to help anticipate fluctuations in entry into emergency departments and downstream services.

4.6. References

Altizer, S., Ostfeld, R.S., and Johnson, P.T.J. *et al.* (2013). Climate change and infectious diseases: from evidence to a predictive framework. *Science*, 341(6145), 514-9.

Axelsen, J.B., Yaari, R., Grenfell, B.T., and Stone, L. (2014). Multiannual forecasting of seasonal influenza dynamics reveals climatic and evolutionary drivers. *Proceedings of the National Academy of Sciences*, 111(26), 9538–9542.

Beard, J.A., Bearden, A., and Striker, R. (2011). Vitamin D and the anti-viral state. *Journal of Clinical Virology*, 50(3), 194–200.

BDSP (2012). Climat, météo et Santé [Online]. Available at: www.bdsp.ehesp.fr/ Base/445731/.

Borella, E., Nesher, G., Israeli, E., and Shoenfeld, Y. (2014). Vitamin D: a new anti-infective agent? *Ann N Y Acad Sci.*, 1317(1),76–83.

Bresee, J. and Hayden, F.G. (2013). Epidemic influenza: responding to the expected but unpredictable. *N Engl J Med*, 368(7), 589–592.

Cannell, J.J., Vieth, R., Umhau, J.C., Holick, M.F., Grant, W.B., and Madronich, S. *et al.* (2006). Epidemic influenza and vitamin D. *Epidemiol Infect.*, 134(6), 1129–40.

Cannell, J.J., Zasloff, M., Garland, C.F., Scragg, R., and Giovannucci, E. (2008). On the epidemiology of influenza. *Virol J.*, 5(1), 29.

Carli, P. (2013). Propositions de recommandations de bonne pratique facilitant l'hospitalisation des patients en provenance des services d'urgences. Report [Online]. Available at: https://www.sante.gouv.fr/IMG/pdf/Rapport_Carli_ 2013_aval_des_urgences.pdf.

Dao, C.N., Kamimoto, L., Nowell, M., Reingold, A., Gershman, K., and Meek, J. *et al.* (2010). Adult hospitalizations for laboratory-positive influenza during the 2005–2006 through 2007–2008 seasons in the United States. *J Infect Dis.*, 202(6), 881–8.

Davies, R., Johnson, D., and Farrow, S. (1975). Planning patient care with a Markov model. *Oper Res Q 1970-1977*, 26(3), 599–607.

Fleming, D.M. and Elliot, A.J. (2007). Lessons from 40 years' surveillance of influenza in England and Wales. *Epidemiol Infect* [Online]. Available at: http://www.journals.cambridge.org/abstract_S0950268807009910. 136(07).

Foster, E.D., Cavanaugh, J.E., Haynes, W.G., Yang, M., Gerke, A.K., and Tang, F. *et al.* (2013). Acute myocardial infarctions, strokes and influenza: seasonal and pandemic effects. *Epidemiol Infect.*, 141(04), 735–44.

Garza, R.C., Basurto-Dávila, R., Ortega-Sanchez, I.R., Carlino, L.O., Meltzer, M.I., and Albalak, R. *et al.* (2013). Effect of winter school breaks on influenza-like illness, Argentina, 2005–2008. *Emerg Infect Dis.*, 19(6), 938–44.

Husson, F., Josse, J., and Pagès J. (XX). Technical report, Agrocampus.

Jhung, M.A., D'Mello, T., Pérez, A., Aragon, D., Bennett, N.M., and Cooper, T. *et al.* (2014). Hospital-onset influenza hospitalizations, United States, 2010–2011. *Am J Infect Control*, 42(1), 7–11.

Lowen, A.C., Mubareka, S., Steel, J., and Palese, P. (2007). Influenza virus transmission is dependent on relative humidity and temperature. *PLoS Pathog.* 3(10), 1470–6.

Lowen, A.C., Steel, J., Mubareka, S., and Palese, P. (2008). High temperature (30 C) blocks aerosol but not contact transmission of influenza virus. *J Virol.*, 82(11), 5650–2.

McLachlan, G. (2004). *Discriminant Analysis and Statistical Pattern Recognition.* John Wiley & Sons.

Mertz, D., Kim, T.H., Johnstone, J., Lam, P.-P., Science, M., Kuster, S.P. *et al.* (2013). Populations at risk for severe or complicated influenza illness: systematic review and meta-analysis. *BMJ*, 347, f5061–f5061.

Noti, J.D., Blachere, F.M., McMillen, C.M., Lindsley, W.G., Kashon, M.L., Slaughter, D.R. *et al.* (2013). High humidity leads to loss of infectious influenza virus from simulated coughs. *PLoS ONE*, 8(2), e57485.

Olson, D.R., Heffernan, R.T., Paladini, M., Konty, K., Weiss, D., and Mostashari, F. (2007). Monitoring the impact of influenza by age: emergency department fever and respiratory complaint surveillance in New York City. *PLoS Med.*, 4(8), e247.

Réseau Sentinelles (2018). Sentinelles [Online]. Available at: https://www.sentiweb.fr/.

Sahu, S.K., Baffour, B., Harper, P.R., Minty, J.H., and Sarran, C. (2014). A hierarchical Bayesian model for improving short-term forecasting of hospital demand by including meteorological information. *J R Stat Soc Ser A Stat Soc.*, 177(1), 39–61.

Schanzer, D.L. and Schwartz, B. (2013). Impact of seasonal and pandemic influenza on emergency department visits, 2003–2010. *Acad Emerg Med Off J Soc Acad Emerg Med.*, 20(4), 388–97.

Serfling, R.E. (1963). Methods for current statistical analysis of excess pneumonia-influenza deaths. *Public Health Rep.*, 78(6), 494.

Shaman, J. and Kohn, M. (2009a). Absolute humidity modulates influenza survival, transmission, and seasonality. *Proc Natl Acad Sci.*, 106(9), 3243–3248.

Shaman, J., Pitzer ,V., Viboud, C., Lipsitch, M., and Grenfell, B.T. (2009b). Absolute humidity and the seasonal onset of influenza in the Continental US. *PLoS Curr*, 1:RRN1138.

Shaman, J., Jeon, C.Y., Giovannucci, E., and Lipsitch, M. (2011). Shortcomings of vitamin D-based model simulations of seasonal influenza. *PLoS ONE.*, 6(6), e20743.

Tamerius, J.D., Shaman, J., Alonso, W.J., Bloom-Feshbach, K., Uejio, C.K., Comrie A. *et al.* (2013). Environmental predictors of seasonal influenza epidemics across temperate and tropical climates. *PLoS Pathog*, 9(3), e1003194.

Te Beest, van Boven, M., Hooiveld, M., van den Dool, C., and Wallinga, J. (2013). Driving factors of influenza transmission in the Netherlands. *Am J Epidemiol*, 178(9), 1469–77.

Tellier, R. (2009). Aerosol transmission of influenza A virus: a review of new studies. *J R Soc Interface.*, 6(Suppl 6), S783–90.

van Noort, S.P., Águas, R., Ballesteros, S., and Gomes, M.G.M. (2012). The role of weather on the relation between influenza and influenza-like illness. *J Theor Biol.*, 298, 131–7.

Willem, L., Van Kerckhove, K., Chao, D.L., Hens, N., and Beutels, P. (2012). A nice day for an infection? Weather conditions and social contact patterns relevant to influenza transmission. *PLoS ONE*, 7(11), e48695.

Zhou, H., Thompson, W.W., Viboud, C.G., Ringholz, C.M., Cheng, P.-Y., Steiner, C. *et al.* (2012). Hospitalizations associated with influenza and respiratory syncytial virus in the United States, 1993–2008. *Clin Infect Dis.*, 54(10), 1427–36.

Part 3

Oncology and Technology

Introduction to Part 3

Cancer remains a mysterious and complicated disease; however, if there is one area where technology can do some good, it is cancer treatment research!

Cancer is an infernal machine that ravages everything in its path, a devastating fire that devours its surroundings and leaves in an arid environment where certain organs struggle to function. The patient suffers this invasion although (which has long since been proven to be unprovoked certain behaviors encourage it): the embers of evil. In the face of this plague, the patient becomes the malleable object that hosts this agony. They are ready to accept anything as long as the barbarity stops forever.

A product of today's world? No, traces of cancer have been found on human skeletal fragments dating from prehistoric times. Hippocrates often made reference to it, as did Henri de Mondeville in the 14th Century to the doctors of Versailles. Cancer's mechanism of attack did not cease to preoccupy the minds until the anatomical design by Laennec in the early 19th Century, followed by the work of Muller and Virchow, which really underpinned cancer's cellular functioning. In 1918, the *Ligue contre le cancer* (league against cancer) was founded in France. Based on the work on X-rays by Wilhelm Rontgen and on radium by Pierre and Marie Curie, X-rays have become the fighter of the impossible and have continued to give hope to the most afflicted.

This idea of a foundation dedicated solely to one type of disease and supporting all the usual care structures shows how complex this disease is.

Complex because more than any other, it ravages the human body and destabilizes its functions, forcing the doctor to treat other diseases at the same time as the mother disease. Complex due to the aggressiveness of possible treatments that, not content to break the evil, also attack the healthy tissues in the manner of General Kutuzov against Napoleon. Complex, finally, due to the moral debacle that it causes, imprisoning the sick between paradise and hell, ready to open the door of Saint Peter to confess their life before the great final leap.

There is no greater strength than combining skills to try to find the best solutions to this disease. The doctor thanks to their acquired knowledge and their permanent manipulation of the human body, and because they also inhabit one, could be confused with a specialized engineer facing this ultra-elaborate machine, but they do not have all the skills to build the elements they need. They could also deal with the spirit that they so often hear in complaint, but they often do not have the time, and perhaps the complacency, to help further. Finally, they could organize all therapeutic care, but this would be at the risk of forgetting a link or getting the wrong patient, so tortuous the process is. Multidisciplinary skills are therefore necessary to optimize the overall care, and many advances in this direction have emerged in the last 10 years. To be continued!

5

Cancer Care Pathway: How Technological Advances are Helping to Address Coordination Challenges

Cancer is always difficult to treat because of the severity of the underlying disease and the complexity of the care management that will follow. Current knowledge, even if it has improved the patients' comfort and survival, has made the care pathways more complex. Technology then becomes a valuable support to help coordinate all the elements of care management.

5.1. Contextual element: current management of cancer patients

In recent years, many advances have emerged in the management of cancer patients in all fields: new drug treatments, more appropriate surgical techniques, less mutilating, more targeted radiotherapy, more precise imaging with the development of functional imaging and better knowledge of tumor biology. At the same time, the development of supportive care has made it possible to better manage the side effects of cancer treatments and symptoms. This includes, for example, pain management, whether related to the spread of the disease or to treatment, and the prevention of nausea and vomiting. Finally, the information provided for patients and their families and friends has clearly improved, becoming a legal obligation, allowing for much more informed involvement in therapeutic choices and follow-ups.

Chapter written by Mario Di Palma.

These therapeutic advances have made it possible to increase the number of patients cured of their cancer and to prolong the survival of other patients, for whom the spread of the disease is not able, at present, to be cured. In a number of cases, we can say that cancer becomes a chronic disease, that is, patients will live with their disease for months or even years.

At the same time, earlier diagnosis and better knowledge of prognostic factors lead to the development of treatments called "adjuvants".

DEFINITION.– "Adjuvant" treatment: a treatment that complements a primary treatment to prevent a risk of local recurrence or metastasis. An adjuvant treatment is a safety treatment. Surgery, chemotherapy (drug or chemical treatment whose vocation in cancer is toxic to cancer cells), radiotherapy (a local-regional cancer treatment method that uses radiation to destroy cancer cells by blocking their ability to multiply), hormone therapy (treatment consists of acting on certain hormones present in the body and promoting the multiplication of certain cancer cells) and immunotherapy (treatment consists of administering substances which will stimulate the body's immune defenses against cancer cells considered as intruders) can be adjuvant treatments in the context of cancer.

For example, after the removal of a tumor, a woman suffering from localized breast cancer should be offered complementary treatments to reduce the risk of recurrence of the disease locally or in the form of remote metastases (in this case, radiotherapy to the breast after the removal of the tumor, chemotherapy and hormone therapy).

These advances in disease-specific care and supportive care have enabled a major breakthrough for patients: the development of outpatient treatment, that is, not requiring one or more nights in hospital, day hospitals and home treatment with oral therapies.

DEFINITION.– Support care: all the care and support needed by sick people throughout the illness. This is done in association with specific cancer treatments that may be implemented. Supportive care responds to needs that may arise during the illness and its aftermath and that may mainly concern the consideration of pain, fatigue, nutritional problems, digestive, respiratory and genitourinary disorders, motor disorders, disabilities and odontological problems. They also concern social difficulties, psychological suffering, disturbances of body image and end-of-life support.

The care management of cancer patients has thus become a model of complexity. It calls upon multiple professional interveners in the community and in the hospital (oncologist doctors for the part of drug treatments (chemotherapy, for example), radiotherapists, surgeons, biologists, radiologists, as well as general practitioners, organ specialists, nurses and pharmacists) who must coordinate to define and apply a common strategy tailored for and with the patient, all the more so as the patient spends less time in hospital.

5.2. Challenges of the development of ambulatory care

These challenges are multiple.

Challenge 1	How can we ensure patient safety with an increased number of healthcare professionals around them?
Challenge 2	How can we ensure that treatments are followed satisfactorily, without undercompliance (reduction in doses or even cessation of treatments) or overcompliance (drugs taken in higher doses than expected)?
Challenge 3	How can the hospital practitioner who is responsible for the treatments they prescribe ensure this responsibility for a patient they see less (as a certain number of patients' direct contact with the doctor prescribing the treatment is limited to consultations at a rate that varies between once every 1 to 3 months depending on the proposed treatment)?
Challenge 4	How can patient information be secured?
Challenge 5	Who can the patient turn to if there is a problem?
Challenge 6	How can information flow between these different health professionals to optimize patient care management?
Challenge 7	And of course, how can we control health costs that are continuously increasing in the context of human resources and professional skills that are limited?

5.3. Care pathway

With the proliferation of care locations and providers, the notion of "care pathways" has emerged, that is, the stages a patient follows from diagnosis through treatment to follow-up. An ill-adapted pathway is synonymous with loss of opportunity for the patient and financial waste for the community.

DEFINITION.– The "optimal care pathway" can be considered as the fact that the patient concerned is "in the right place, at the right time, with the right person", that is, avoiding a patient coming to the hospital or being hospitalized when it is not necessary or, conversely, avoiding leaving a patient at home when an intervention is necessary.

All these elements (the number of patients increasing, outpatient treatment, the increasing complexity of care, the increase in the number of caregivers) require the development of an effective inter-functional coordination in patient care. In this respect, it should be stressed that oncology is an example of complex management, and also that the current thinking in this field is perfectly transposable for any chronic pathology.

5.4. Inter-functional coordination

This inter-functional coordination can be performed by a variety of people depending on the patient's situation, pathology, level of autonomy, environment and so on. Currently, this role is often assigned to specialized nurses called "coordination nurses". The objective is to maintain the link between the patient, their entourage and their various health professionals, thus making it possible to direct and guide the patient towards the expertise they need, which is not necessarily found in the hospital. These coordinating nurses will also liaise with independent health professionals to enable them to interact as effectively as possible with their patients.

A few years ago, the National Cancer Institute (NCI) conducted the first experiment at the national level concerning cancer coordination nurses, which made it possible to show the relevance of this function and to define its scope. The second part of this experiment is currently under way and should make it possible to model inter-functional coordination. As an example, for more than 10 years we have been developing a specific coordination service (external care coordination) for patients at home who require complex care at the Gustave Roussy Institute in France.

This coordination service links the patient at home and the professionals working with him/her (attending physician, pharmacist, nurse, provider), either systematically at a predefined pace or in the event of a particular problem. More than 3,000 patients were treated last year by this structure. It is interesting to note that almost half of the calls received by a coordinating nurse were from independent health professionals.

5.5. Patient-reported outcome (PRO) information

Treatments, especially medications, used in oncology can sometimes cause serious side effects. The balance between the benefits of the treatment and the side effects must therefore be continuously assessed. For this purpose, it is necessary to evaluate these side effects in the best possible timely manner and in the most relevant and accurate way. In most cases, this is done during consultations with the oncologist, often on a monthly basis.

When the coordinating nurse takes action with the patient, the overall contact with the nurse, which until now has essentially been by telephone, can increase in frequency. Several studies have shown that when the patient reported the information in notebooks, which were filled in regularly, this information was more relevant and much closer to reality. This is what we find in the literature under the term, PRO ("patient-reported outcome").

5.6. Limits of coordination and the place of communication tools

In an ideal world, each patient should be followed up by a coordination nurse they could turn to in the event of a problem, with a logbook (PRO) to report and identify all the problems they encounter, whether symptoms related to the disease or side effects of the received treatments. Obviously, there is very quickly a problem of time, availability of health professionals and management of a mass of information that needs to be processed to perform a satisfactory analysis that leads to decisions adapted for each patient. It is in this framework that modern communication and telemedicine tools create much interest.

5.7. Specifications for communication tools?

Prerequisite: first, the data transmission must be secure, guaranteeing the confidentiality of the information. This is the role of the *Commission nationale informatique et liberté* (*CNIL*, the French national commission for informatics and liberties), which must validate the tools in question.

First condition: it is a question of transmitting information in real time and issuing an alert if need be. It is certain that screening and, considering complications (whether related to the disease or to treatments) allows for much more appropriate management and reduces the risks for patients. Let us take a simple example: a patient who has difficulty eating and drinking and also has diarrhea problems is at risk of dehydration. If this dehydration is quickly identified, the consequences will be limited for patients and management will be relatively simple (in the form of infusions, for example, which can even be organized at home if necessary). If dehydration develops, it can have serious or even vital consequences for the patient, and the need to hospitalize the patient may arise.

Second condition: the information transmitted must be exhaustive and the tools used must be reliable. The dissemination of information must be targeted at people who are actually able to react. The information must also generate feedback to reassure, advise or alert the patient, depending on the situation.

Third condition: above all, the information must be processed. Potentially, the connected tools can generate a large amount of data. Raw transmission of this data would be inefficient and unnecessary. The tool must therefore include a sorting and analysis function to only transmit the alert when relevant and according to the thresholds that must be defined according to the patients.

5.8. What are the guidelines for connected tools?

PROBLEM.– Which information is worth transmitting and which has an interest in patient care?

They are potentially multiple. On the one hand, there is all the information that the patient can generate on their own and on a regular basis, for example, the level of pain according to a numerical scale or the degree of fatigue. It can also be interesting to transmit data in an open way: the patient may report unusual symptoms that may feel like an abnormal shortness of breath or a headache.

SOLUTION.– Internet-connected tools can record and transmit, without patient intervention, a number of parameters that can be important for monitoring the patient, such as temperature, blood pressure, heart rate and weight.

In the long run, it is quite likely that the patient will be able to carry out fairly complete biological analyses at home with a simple drop of blood, beyond what is currently done with sugar levels.

Of course, it is out of the question to record all possible data continuously and for all patients. Parameterization should be possible on a case-by-case basis and adaptable according to the patient's progress.

CONDITION OF APPLICATION.– The confidentiality of the reported data must be guaranteed. There is therefore an issue of securing information in its collection and transmission. It is also necessary to explain to the patient the interest of these tools: the objective is obviously not to interfere in their private life, but it is advisable to limit the recorded data to what is really useful for their follow-up and security.

Today, a majority of patients are ready to accept these communication tools. On a survey we conducted at the Gustave Roussy Institute (Girault *et al.* 2015), 85% of the patients had a smartphone, a tablet or a computer and would be able, provided of course that the system was secure, to use them for their follow-up.

Most studies show that patients are prepared to a certain extent, of course taking into account particular cases (fatigue, whether physical or mental, vigilance problems, confusion) to inform themselves a certain number of data to make this link with coordination structures.

5.9. Internet-connected tools and information systems

Another challenge is to integrate the data generated by Internet-connected tools and by the patient into the patient's digital file, whether it is the hospital patient record or the general practitioner patient record. There is an issue of system interoperability. It is indeed necessary that Internet-connected tools can communicate via their own interconnections and also with the information systems of health professionals. This is not necessarily in the culture of the developers of the tools in question. Several experiments are currently underway within the framework of *Terr-E Santé* to organize the interconnection of information systems.

5.10. Digital tools and the human element

Today, many applications exist as well as connected platforms allowing information to be transmitted in complete security. Internet-connected tools are available to meet most of the previously described parameters. Finally, other tools using artificial intelligence are being developed.

The risk is to initiate reflection from the tool and to be somehow fascinated by the technology without sufficiently asking the question of relevance and especially integration in a process of empowerment. For example, the Internet-connected scale is surely something very useful to daily motivate someone to lose weight, but for a patient's follow-up, the weight tracking alone does not really have much interest.

Another major problem, beyond the processing and sorting of information, concerns the contextualization of this information.

For example, a patient indicates a pain level of 7/10 (which corresponds to significant pain) and if they did not have pain until then, it is advisable to intervene quickly; if they had a prior even more intense pain, at 10/10, with a treatment that was initiated, it may mean that the treatment in question is beginning to be effective. There is therefore an indispensable analysis of the patient's context, which necessarily requires human intervention that will consider the specific situation of the patient concerned.

5.11. Coordinated follow-up, Internet-connected medicine and impact on patient survival

Several recent studies show that organized patient follow-ups not only improve a patients' quality of life but also their survival.

We can quote the work of Dr. Ethan Basch (Basch *et al.* 2017), which concerns patients undergoing chemotherapy followed in a usual way or with an application allowing them to report their symptoms, with an improvement in their survival.

This is also the case with the study recently presented by Dr Fabrice Denis (Denis *et al.* 2017), which shows that in patients treated for localized lung cancer, follow-up by an application informed by the patients makes it possible to detect a possible recurrence of the disease much earlier and thus can trigger an adapted management earlier too, which results in an improvement in their survival.

These results also open the question of the financing of Internet-connected medicine, human tools and interventions.

5.12. Conclusion

Oncology is a complex management model for patients with multiple stakeholders and therapeutic advances that allow increasingly prolonged outpatient treatment and survival. These care pathways require coordination to be optimized. Today, the technological tools to facilitate patient follow-ups exist. The integration of these tools in the service of coordination structures will not only significantly improve the future of patients treated for cancer but will also optimize human and financial resources in the health field. This is a real public health issue.

5.13. References

Basch, E. *et al.* (2017). Overall survival results of a trial assessing patient-reported outcomes for symptom monitoring during routine cancer treatment. *JAMA*, 318(2), 197–198.

Denis, F. *et al.* (2017). Randomized trial comparing a web-mediated follow-up with routine surveillance in lung cancer patients. *Journal of the National Cancer Institute*, 109.10.1093/jnci/djx029.

Girault, A. *et al.* (2015). Internet-based technologies to improve cancer carecoordination: current use and attitudes among cancer patients. *Eur J of Cancer*.

Optimization Issues in Chemotherapy Delivery

6.1. Background presentation

Cancer chemotherapy involves a group of substances that prevent or even kill cancer cells. The manufacture of cancer chemotherapy agents involves the handling of toxic products and hence it is a complex and expensive process (Maraninchi *et al.* 2016).

Nevertheless, the efficacy of cancer chemotherapy is now established without controversy. It allows us to obtain interesting results in certain pathologies. The active products of chemotherapy treatments are so-called cytotoxic (toxic to the cell) drugs. These toxic products have side effects for the patient, and they also pose risks for the people handling them. These risks related to the preparation and handling of the toxic products are significant. On the other hand, there are many manufacturing constraints to consider: each preparation has a dosage adapted to each patient; the preparation must be ready in time to be administered to the patient and the sterility of the preparation must be guaranteed while protecting the personnel carrying out the preparation.

The *Centre hospitalier régional et universitaire (CHRU) de Tours* occupies a privileged place in the Centre-Val de Loire region in France. At the Bretonneau Hospital, where the *Centre régional de cancérologie Henry S.*

Chapter written by Jean-Charles BILLAUT, Virginie ANDRÉ, Yannick KERGOSIEN and Jean-François TOURNAMILLE

Kaplan (oncology center) specializing in oncohematology is located, all types of cancer are treated. This center has a pharmaceutical unit called the *Unité de biopharmacie clinique oncologique* (UBCO, the oncology clinical biopharmacy unit), certified ISO 9001, which produces on average 25,000 preparations each year for oncohematology specialties (Datalogic success stories, Aubert *et al.* 2009).

DEFINITION.– ISO 9001 standard (International organization for standardization 9001): this is a standard established to take into account the quality management for a given product. It is based on a certain amount of information provided in a document by the manufacturer and approved by a recognized organization. This document determines the rules, guidelines and characteristics of a product that guarantee an optimum level of order and safety when using the product.

In order to improve the chemotherapy production process and the quality of patient care, several optimization issues have been identified. The following three issues are presented in this chapter (Billaut 2014): the problem of optimizing the production of preparations, the problem of optimizing the consideration of residues and, finally, the problem of optimizing distribution. For each of these, we present an example, a mathematical model of a case and a discussion on the general case.

6.2. Production planning issues

We begin by describing the production environment of a unit like UBCO.

A cancer chemotherapy preparation unit is a controlled atmosphere area. Each preparation device present within the unit is a completely enclosed system called an isolator.

Several types of isolators may exist; a UBCO isolator is shown in Figure 6.1 This device consists of a first part, called the sterilizer, which can contain up to 12 baskets placed on a central rail. Each basket corresponds to a preparation to be made, intended for a patient. All the baskets placed in the sterilizer at the same time are called a "batch". This then connects with the sterile isolation chamber, where two preparers can work face to face at the same time (on some devices, they are side by side, and on others, more than two preparing stations are available). The preparations made are evacuated by an airlock, and the waste is evacuated in containers provided for this purpose, placed under the isolation chamber (they are to be incinerated).

The preparation procedures are carried out in accordance with the prescriptions of oncologists.

The different manufacturing phases are described in Figure 6.1.

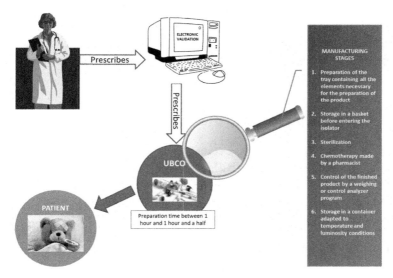

Figure 6.1. *Chemotherapy implementation steps*

The following photographs show an isolator and a sterilizer.

Figure 6.2. *Isolator (left) and part of the sterilizer carrying the baskets (right). For a color version of the figure, please see www.iste.co.uk/sarazin/health.zip*

This production line is covered by additional controls, arriving at different stages.

All these activities have been computerized, and two software programs have been implemented at UBCO to ensure complete traceability of manufacturing (from prescription to patient administration) and to plan activities on a daily basis [6.19]. A total of 10 steps are outlined for each chemotherapy preparation. These are mainly the steps related to the preparation and control of the tray (bags, cytotoxic product, etc.), the sterilization of the elements (equipment number or isolator and load cycle), the preparation (dosages in a controlled environment) and various controls (visual, weighing or analytical).

6.3. Modeling the scheduling problem

It is possible to propose a complete model of the chemotherapy preparation production system (Billaut *et al.* 2014). Such a model makes it possible to study the behavior of the system in the event of an increase in the workload or in the event of a hazard.

However, a complete model is not essential if the objective is to propose a tool to guide the daily production of the service. To provide an interactive decision support tool, a "reduced" model of the production system is sufficient (Mazier *et al.* 2007; Mazier *et al.* 2010) to the extent that some decisions are not made by the system but are deliberately left to a decision-maker.

6.3.1. *Complete model*

In this part, a more generic vocabulary is adopted, which departs from the field of application and approaches scheduling problems (operational research field). In particular, "job" is the making of a preparation; "desired delivery date" is the date by which the preparation must have reached the patient and "machine" is one of the two production lines associated with an isolator (that is, a dispensing pharmacist).

The problem is to schedule a set J of n jobs. Each job J_j is associated with a runtime denoted by p_j (which varies according to the preparations) as well as a desired delivery date (or due date) denoted by d_j. Each job J_j is also associated with a start date at the earliest noted r_j, which corresponds to the approximate date on which validation of the doctor's prescription is expected (in Figure 6.3, the end date of the doctor's visit corresponds to this date r_j). The sterilization time is the same for all jobs, and the inspection time is also the same.

An isolator, that is, a sterilizer, and the two operators associated with it, can be considered as a small two-step production workshop with a "max-batch"-type machine with finished capacity on the first stage and two machines parallel to the second stage. Therefore, there are as many small workshops of this type as isolators. All the jobs have the same production range, which consists of passing through a workshop of this type, then finishing with a single, common machine, which is the one that carries out the control. A Gantt diagram representing the progress of some jobs is shown in Figure 6.3.

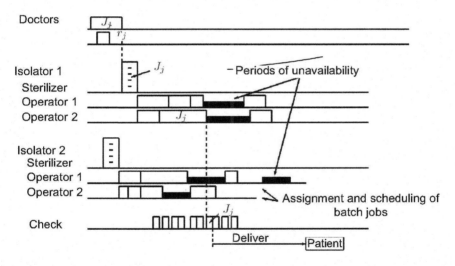

Figure 6.3. *Overall model of the preparation production workshop*

The problem consists of determining, for each job, which isolator is assigned to it and on which date sterilization starts (same date for all job batches); which isolator machine it is then assigned to and on which date it is performed and, finally, on which date the control on the last machine takes place.

Over a time frame of 1 day, UBCO makes about 150 preparations. Given the complexity of this type of production facility, the use of an exact (optimal) method to resolve the entire day's planning cannot be done in a timely manner. To solve this problem, it is thus necessary to have recourse to speed up methods, called "approximates" because they do not guarantee one will find the optimal solution (Tabu method, genetic algorithm, etc.).

6.3.2. *Scale model used*

To propose an interactive decision support method, the planning problem was broken down into the following three phases, naturally leading to a simplification of the workshop model:

1) at each decision moment (approximately every 2 minutes, requests are made to know the new validated prescriptions), an assignment to an isolator of each preparation to be made is proposed;

2) the decision-maker validates certain proposed assignments and sets up his/her own batches to initiate sterilization;

3) when the batches to be sterilized are validated by the decision-maker, assignment and scheduling of the preparations for each compounding pharmacist are proposed.

Assignment to isolators. At a given moment, we denote by J_1 all the jobs to be done, already assigned to an isolator. J_2 is the set of new jobs sent to the department and not assigned to an isolator. Each machine is associated with unavailability periods (allowing for staff arrival times and breaks). Note that two machines are associated with the same isolator.

The procedure involves sorting the jobs of $J_1 \cup J_2$ in the order of the increasing desired delivery dates. Then, the jobs are taken in that order and assigned to the isolator that contains the machine that allows the earliest completion of the job. However, J_1's job remains assigned to the same isolator.

These assignments are proposed to decision-makers in the form of a list (see Figure 6.4). Among all the jobs assigned to an isolator, a decision-maker selects the jobs he/she wants to produce and then starts the calculation that will insert them into the schedule.

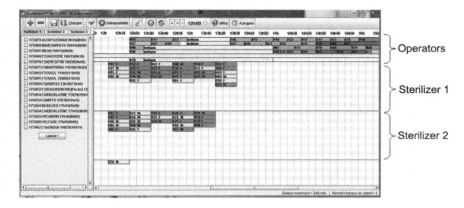

Figure 6.4. *Screenshot showing (left) the list of jobs assigned to isolator 1*

We note that the baskets corresponding to the unselected jobs are stored in a temporary zone, waiting to pass through the isolator (see Figure 6.5).

Figure 6.5. *Storage area for baskets ready for sterilization. For a color version of the figure, please see www.iste.co.uk/sarazin/health.zip*

Scheduling and assignment of tasks. The assignment and scheduling within an isolator is modeled by the following linear program using time-indexed variables [6.10]. We denote by n the number of jobs in the batch to be scheduled and $x_{j,t}$ a binary variable with value 1 if the job J_j is running on the date t and 0 otherwise ($1 \leq j \leq n$ and $0 \leq t \leq H$) with H the duration of a working day broken into steps of 5 minutes (an amplitude of 10 working hours leads to a value of $H = 120$).

The objective function under consideration aims to reduce patient waiting times. If we denote by T_j late delivery of the preparation J_j, this is the maximum delay function denoted by $T_{max} \geq 0$, defined as $T_{max} \geq max_{1 \leq j \leq n} T_j$, which must be minimized.

The constraints are as follows.

Each job must be fully completed, that is, $\forall j, j \in \{1, \dots, n\}$:

$$\sum_{t=r_j}^{H} x_{j,t} = p_j \qquad [6.1]$$

At any given moment, there can be no more than m_t job in progress, that is, $\forall t, t \in \{0, \dots, H\}$:

$$\sum_{j=1}^{n} x_{j,t} \leq m_t \qquad [6.2]$$

with m_t the number of machines available at the moment t (this allows us to take into account the unavailability of the preparers, known in advance).

Finally, a job cannot be preempted, it must be done in one go. We have $\forall j, j \in \{1, \dots, n\}$ et $\forall t, t \in \{0, \dots, H\}$:

$$p_j(x_{j,t} - x_{j,t+1}) + \sum_{t'=t+2}^{H} x_{j,t'} \leq p_j \qquad [6.3]$$

This constraint reflects the fact that as soon as $(x_{j,t} - x_{j,t+1})$ becomes 1, in other words, as soon as the job J_j stops, jobs can no longer be performed at any date $t' \geq t + 2$ (the p_j are simplified on each side). In other words, jobs are stopped only once, which prohibits pre-emption.

The objective function takes its value thanks to the following constraints: $\forall j, j \in \{1, \dots, n\}$ and $\forall t, t \in \{0, \dots, H\}$:

$$T_{max} \geq t \times x_{j,t} - d_j \qquad [6.4]$$

to the extent that the end date of the J_j is the highest value of $t \times x_{j,t}$, its delay is the highest value of $t \times x_{j,t} - d_j$.

The model is written as:

MIN T_{max}

s.c. (1), (2), (3), (4)

$T_{max} \geq 0$

$x_{j,t} \in \{0,1\}, \quad \forall j \in \{1, \dots, n\}, \quad \forall t \in \{0, \dots, H\}$

This model has nH binary variables and O(nH) constraints. With a relatively small value of n (n $\leq$ 12) and T = 120, that is a maximum of 1,440 variables.

Once this linear program has been solved using a solver (that is, GLPK and Gurobi), the assignment of the job to the machines must be carried out (we know it exists, the constraints guarantee it, but the model does not provide the assignments). By considering each job as a fixed time interval, the definition of a job assignment to machines can be solved by a bicolor problem in an interval graph, which can be solved in polynomial time [6.21].

6.3.3. *Implementation and impact*

The thus obtained solution can very easily be implemented. The machine assignment indicates at the time of entry into the sterilizer on which side of the rail the basket corresponding to the job should be placed and the sequences on each machine indicate in which order the baskets should be placed.

The IT solution was implemented through PLANIF software, which enabled UBCO to switch from a manual management mode for sterilization launches to a tool-guided launch, with a good level of readability on the workload of the day (Tournamille *et al.* 2007).

The PLANIF tool allows the production of jobs to be smoothed over time according to needs, which has had the effect of significantly reducing waiting times for all departments, with sterilization schedules different from those used without the application. From the internal point of view of the operation of the UBCO, the planning made it possible to avoid filling the isolators with non-emergency preparations, to have a margin of safety in the event of an emergency treatment and to find a significant space saving at the level of the working surface. The average sterilizer load is about seven baskets at a time (Aubert 2009, p. 30).

6.4. Problem linked to the consideration of residues

In this section, we study the circuit used by cytotoxic products necessary for chemotherapy.

6.4.1. *Presentation of the problem*

The bottles of cytotoxic active ingredients, referred to here as raw materials, are stored in specially designed refrigerators. As long as they are not opened or reconstituted, they are considered non-perishable and in infinite quantities. It is assumed that a stock management system is in place to prevent shortages.

Once a bottle is taken out of the refrigerator, it is placed in a basket to make a preparation. In the isolator, the bottle is opened by the dispensing pharmacist and possibly shared by all the preparations that require it. In other words, the same bottles can be used in several preparations, if they are sterilized at the same time.

Once the bottle is opened, the product is activated, and it acquires an expiry date, which depends on the nature of the contents. The product remaining in the bottle after the quantity necessary for preparation has been taken, which is called residue.

Once the batch is finished, if there is still material left in the bottle, it is put back in the refrigerator until it is next used or until it is disposed of because the expiry date has passed.

The raw materials used for cancer treatments have several characteristics:

– They are *very expensive*. Cancer drug prices are described as "exorbitant" and even "unfair" (Maraninchi 2016). For example, Keytruda, used for certain lung cancers and known to have removed Jimmy Carter's tumor, aged 91, is sold at a price of 100,000 euros per year in the United States (Delchaux 2016: about 9,000 US dollars for four bottles of 50 mg). In (Maraninchi 2016), it is stated that "American cancer specialists have expressed their concerns about the prices of these innovations, moving to see them rise from 10,000 to more than 120,000 dollars per patient per year in fifteen years". These costs are too high for social security (Paillé 2016).

– They are *unstable*. Once reconstituted, a cytotoxic product has a very limited shelf life. In the same way as an antibiotic, once reconstituted, it must be stored in optimal conditions (refrigerator) and consumed relatively quickly, before losing its properties. Data on the physicochemical stability of injectable anti-cancer drugs are not readily available. Studies have been conducted on this stability (Respaud 2011) to avoid their waste. On the other hand, once the preparation has been made, it must be administered to the patient before a certain period of time has elapsed. There is therefore both an expiry period for the residue contained in the bottle and an expiry period for the preparation once it has been made. For example, Eloxatin (used against colon cancer) retains its physicochemical properties for 24 hours after reconstitution and the infusion solution should be used immediately. Oxaliplatin (used against cancer of the large intestine) retains its physicochemical properties for 48 hours after reconstitution and the infusion solution should be used immediately. Dacarbazine (used for the treatment of metastatic malignant melanoma) has a stability of 1 hour after reconstitution, and the stability of the diluted infusion solution is 30 minutes.

However, it should be noted that drug package inserts do not contain preservatives and hence are not intended for multiple use. In other words, the recommendations are to throw away any bottles that are not fully used, and therefore to not have any residues. In Respaud's study (Respaud 2011), it is indicated that a fine management of the residues allowed a saving of approximately 10% of the annual budget of injectable anti-cancer drugs, which represents for 1 year a sum of approximately 750,000 euros.

We place ourselves in this context and propose to optimize the use of the bottles, rather than systematically discarding their contents.

We start by showing the difficulty of the problem in a very simplified environment. We then present a model of the problem in the global environment.

6.4.2. *Special case: one machine and one product*

Let us consider a production workshop composed of a single machine. It is assumed that all the jobs to be performed (which are independent tasks) consume a certain quantity of the same product. Therefore, we only have one anti-cancer drug to use, the same for all the jobs. We need to schedule a set J of n jobs. Every job J_j is characterized by a noted execution time p_j, a desired end date noted d_j and a quantity b_j, the consumption of the product. We know the price of a noted bottle for the product W, the volume of a noted bottle V (it is considered that there is only one possible capacity) and the shelf life of the product in the bottle after reconstitution, denoted by T.

Without loss of generality, it is considered that $b_j \leq V$, $\forall j, j \in \{1, \dots, n\}$. It is assumed that the time to deliver the preparation to the patient is much longer than the stability of the preparation, so this time is not of concern, but the stability of the product in the bottle after reconstitution is.

It is possible to define new objective functions associated with the consideration of residues (Billaut 2011):

– the first is linked to the economic aspect, in other words, to the costs of discarded products;

– the second is linked to the ecological aspect, in other words, to the quantity of products thrown away.

Since we place ourselves in a static context (the number of jobs is known and fixed), we know precisely the minimum quantity of products to use. This quantity is equal to $B = \sum_{j=1}^{n} b_j$. The minimum number of bottles to open is therefore equal to $F^{min} = \left\lceil \frac{B}{V} \right\rceil$. The mandatory minimum loss is $Q^{min} = F^{min} \times V - B$. If we consider a given scheduling σ, which requires the opening of $F(\sigma)$ bottles (on an $F(\sigma) \geq F^{min}$), then the quantity of product lost is therefore equal to

$$Q(\sigma) = \left(F(\sigma) - F^{min}\right) \times V + Q^{min}$$

The cost of the σ solution is equal to

$$K(\sigma) = W \times F(\sigma)$$

The quantities V, F^{min}, Q^{min} and W are constant (in our particular case, where there is only one type of bottle).

The problem of finding a solution σ that minimizes the quantity of product lost $Q(\sigma)$ is therefore equivalent to the problem of finding a solution that minimizes the cost of open bottles $K(\sigma)$ and amounts to minimizing the number of open bottles $F(\sigma)$.

In order to take into account the desired end dates and not to degrade the solution too much, a tolerance threshold on the value of the greatest delay can be defined. For example, the value of the greatest delay must be less than or equal to a certain value ε. If we denote by C_j the end date of the jobs J_j, then we define the delay of J_j with the variable $T_j = \max(0, C_j - d_j)$ (as before) and the maximum delay per $T_{max} = max_{1 \leq j \leq n} T_j$. We have the following constraint:

$$T_{max} \leq \varepsilon$$

equivalent to

$$C_j \leq d_j + \varepsilon, \quad \forall j \in \{1, ..., n\}$$

EXAMPLE.– Let us consider a set with $n = 6$ jobs with $V = 10$, $T = 10$ and the following data:

j	1	2	3	4	5	6
p_j	8	7	5	2	3	5
b_j	2	4	5	8	6	5
d_j	1	1	2	2	2	3
	5	7	0	2	5	0

The minimum number of bottles to open is $F^{min} = 3$ because $\sum b_j = 30$ and $V = 10$.

One solution to the problem can be represented by a Gantt chart where each job takes one dimension for time, the other for its resource consumption. Two jobs using the same bottle combine the two dimensions, time and volume, and are therefore represented by making the upper right corner of a job coincide with the lower left corner of the job that follows it.

The optimal solution for the T_{max} is the solution given by the EDD (earliest due date first or increasing d_j order), or the sequence $\sigma = (J_1, J_2, J_3, J_4, J_5, J_6)$ presented in Figure 6.6 (a symbol indicates the opening of a new bottle). The greatest delay is 0.

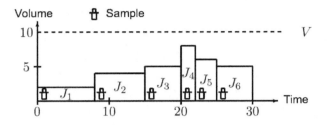

Figure 6.6. *Gantt diagram for the sequence*
$\sigma = (J\_1, J\_2, J\_3, J\_4, J\_5\ [\!, J]\!\!\_6)$

In this solution, jobs J_1 and J_2 cannot use the same bottle because $p_1 + p_2 > T$, just like jobs J_2 and J_3 (because $p_2 + p_3 > T$). Jobs J_3 and J_4 cannot use the same bottle because $b_3 + b_4 > V$, just like jobs J_4 and J_5 (because $b_4 + b_5 > V$). Finally, jobs J_5 and J_6 cannot use the same bottle because $b_5 + b_6 > V$. This solution, for which the greatest delay is equal to 0, requires the opening of $F(\sigma) = 6$ bottles, one per job.

Now, we consider the sequence $\sigma = (J_1, J_4, J_2, J_5, J_3, J_6)$ depicted in Figure 6.7.

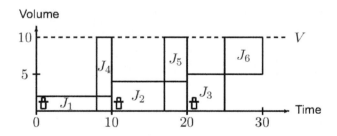

Figure 6.7. *Gantt diagram for the sequence*
$\sigma = (J\_1, J\_4, J\_2, J\_5, J\_3\ [\!, J]\!\!\_6)$

In this solution, jobs J_1 and J_4 can use the same bottle, the same for J_2 and J_5, and also for J_3 and J_6. In total, only $F(\sigma) = 3$ bottles are used. We note that $F(\sigma) = F^{min}$, which means that the solution is optimal for this criterion (you cannot find a solution with fewer bottles). However, the value of T_{max} is no longer equal to 0 but equal to 5 (job J_3 ends on date 25 when it is due on date 20). The solution is therefore degraded in terms of maximum delay but improved in terms of the number of bottles used. This shows that the two criteria are in conflict. We refer to T'Kindt (2006) for a general presentation of multi-criteria scheduling problems.

COMMENT.– If it is assumed that a chemotherapy preparation can be made from several bottles, then the model should be adapted accordingly. In this case, for the example shown in Figure 6.6, the J_6 would not need a new bottle.

The *linear programming model* problem can be modeled as a linear integer program. We denote by u_k a binary variable equal to 1 if the bottle K is used (Billaut 2015). If all jobs have a due date equal to the sum of the durations (that is, ignoring the constraint on the largest delay), then the problem is exactly the problem called "two-constraint bin packing", also called the "vector packing problem" (see, for example, Alves 2014). the problem can be modeled as a linear integer model. We cal u_k a binary variable equal to 1 if the bottle k (we also say the _bin_ k) is used, and 0 otherwise. We call $y_{j,k}$ a binary variable equal to 1 if job J_j is assigned to the bottle k, and 0 otherwise.

It is assumed that the jobs are numbered in EDD order.

We try to minimize the number of bottles used, let $\sum_{k=1}^{n} u_k$.

Each job must necessarily be assigned to a bottle, in other words, $\forall j \in \{1, \dots, n\}$, we have:

$$\sum_{k=1}^{n} y_{j,k} = 1 \tag{6.5}$$

The total duration of the job in a bottle may not exceed the time limit T, that is, $\forall k \in \{1, \dots, n\}$:

$$\sum_{k=1}^{n} p_j y_{j,k} \leq T \times u_k \tag{6.6}$$

The total consumption of the job in a bottle may not exceed volume V, that is, $\forall k \in \{1, \dots, n\}$:

$$\sum_{j=1}^{n} b_j y_{j,k} \leq V \times u_k \tag{6.7}$$

If we denote by ε the value that the greatest delay cannot exceed, we then have $\forall j \in \{1, \ldots, n\}$ and $\forall k \in \{1, \ldots, n\}$:

$$\sum_{h=1}^{k-1}\sum_{i=1}^{n} p_i y_{i,h} + \sum_{i=1}^{j} p_i y_{i,k} \leq d_j + \varepsilon + M(1 - y_{j,k}) \qquad [6.8]$$

The expression $\sum_{h=1}^{k-1}\sum_{i=1}^{n} p_i y_{i,h}$ gives the sum of the duration of the job in the $k - 1$ first bottles (bins). Furthermore, $\sum_{i=1}^{j} p_i y_{i,k}$; in other words, the sum of the duration of the job precedes J_j in the bin (the jobs are numbered according to EDD, so this order takes precedent within a bin) plus J_j. The obtained end date of the job is thus J_j, which must be less than or equal to $d_j + \varepsilon$. This constraint should only be satisfied if J_j is in the bin k, hence the presence of $M(1 - y_{j,k})$.

To eliminate symmetries, the following constraints are added to ensure that the bottles are used in the order of their increasing numbering, $\forall k \in \{1, \ldots, n\}$:

$$u_{k+1} \leq u_k \qquad [6.9]$$

The model is written as:

MIN $\sum_{k=1}^{n} u_k$

$s.c. [6.5, 6.6, 6.7, 6.8, 6.9]$

$u_k \in \{0,1\}, \quad \forall k \in \{1, \ldots, n\}$

$y_{j,k} \in \{0,1\}, \quad \forall j \in \{1, \ldots, n\}, \quad \forall k \in \{1, \ldots, n\}$

This model includes $n(n + 1)$ binary variables and $n^2 + 4n$ constraints.

6.4.3. General case

For a real implementation, the circuit of the cytotoxic products should be studied finely. A non-reconstituted bottle is taken out of stock, a refrigerator, and placed in the basket in which it is used. As long as it is not opened and reconstituted, the product is considered viable and can be kept for a sufficiently long period of time. The bottle is opened in the isolator, reconstituted and possibly shared between all the preparations that require it. If a preparation requiring the bottle is expected within a short time, it may remain in the isolator for some time. Otherwise, it comes out and is placed back in the refrigerator. This time, its life is limited and it cannot exceed a certain time (see Figure 6.8).

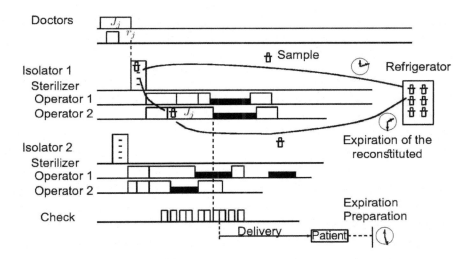

Figure 6.8. *Comprehensive model of the preparation production workshop with bottle circuit* (Billaut 2011)

For each product, we know the volume of the containers, specific to each product, the price per unit volume and the shelf life once reconstituted.

A linear programming model can be developed based on the mathematical model in section 2.1 and on the mathematical model of this section generalized to several types of bottles. Such a model requires a very large number of binary variables and a very large number of constraints, which makes the model unusable for practical use. On the other hand, the development of a metaheuristic method approach is justified.

An interesting way to better optimize the management of residues is to go through optimization at two levels. Indeed, the shelf life of products in open bottles often exceeds the day. It is therefore necessary to broaden the working time frame: a time frame of 30 days is undoubtedly sufficient for a relevant study. On such a time frame, given the complexity of the problem, it is then possible to develop planning software that will study the best distribution of preparations over a 30-day time frame, to reduce product losses as much as possible, given their stability. The implementation of such a production plan then comes down to coordinating appointments with patients – when protocols allow this, of course – and this is with the various oncology services. Then, at a second level, it is possible to schedule the preparations by the day, while optimizing the use of the residues. The implementation of such a procedure requires prior acceptance by all the services concerned.

6.5. Consideration of distribution

Due to the high volatility of the drugs prepared and in order to minimize patient waiting, the distribution of preparations is an important aspect of the problem. The problem has two specificities: transport times are not negligible compared to production times and it is not possible to indefinitely store the preparations made pending their distribution. It is therefore necessary to coordinate production and distribution. This aspect of the problem is discussed here and is referred to in the literature as "*integrated production and distribution*".

6.5.1. *Presentation of the problem*

At the CHRU in Tours, distribution takes place at three different hospital sites, including two remote from the production center and that require a vehicle.

Various articles considering this problem can be found in the literature (Bilgen 2004; Chen 2004). Much of the literature is devoted to problems that arise at a strategic level. We are here at an operational level and many fewer items address these problems (see for example Kergosien 2017; Viergutz *et al*. 2014; Ullrich 2013).

The resolution of this problem requires the resolution of three interrelated sub-problems:

PROBLEM 1	Production scheduling presented in the previous sections.
PROBLEM 2	Creation of "batch tours" (different from sterilization batches), which consist of determining which preparations will be delivered together in the same round, that is, assigned to the same batch/round. Once a batch is known, a departure date for the round is associated with each batch. This date corresponds to the end time of production (that is, control) of the last batch preparation of the round.
PROBLEM 3	Design of each round. This second problem consists in determining the order in which deliveries are made within each batch. This problem is also called the "vehicle routing problem" in the literature (Desrosiers 1995) and has been the subject of numerous studies because of its complexity as the number of deliverables increases.

The last two problems are usually solved together because the choice of preparations composing the batch has a direct impact on the duration of the round (delivery time and delivery date of each preparation). Distribution decisions are difficult to make and have a significant impact on delivery dates. Here are two examples:

– *Example 1:* if the decision-maker adopts a strategy that consists of only delivering a few preparations in the same round, so that the delivery person must return more often to the place of production, then they will be led to make many round trips that are probably useless and thus the number of chemotherapies ready to be delivered is likely to accumulate very quickly. This effect will cause a significant increase in delivery dates, especially for chemotherapies from the last rounds.

– *Example 2:* if the decision-maker adopts a strategy of delivering all chemotherapies ready to be delivered (produced and stored) without waiting, then the delivery person's rounds may become longer and longer and ultimately have the same effect as before. For example, if two preparations intended for two patients on the same ward, 15 minutes from the production site for example, are completed at 9:55 am for the first and 10:05 am for the second. Suppose the delivery person is back to start a new round at 10:00 am. Then they will take the first chemotherapy but not the second. They will have to come back to the same department on their next round, whereas if they had waited 5 minutes, then they could have delivered both chemotherapies at the same time and therefore saved time overall.

On the other hand, it is difficult to estimate the time the delivery person must wait for their next round. To wait too long would cause too many preparations to be accumulated and ready for delivery. Distribution decisions are therefore very strongly linked to production decisions. Indeed, producing preparations in a new order impacts the end dates of production and therefore requires a modification of batches and rounds. These production decisions must therefore also take into account the delivery locations of chemotherapies and delivery dates. Finally, an important problem constraint adds complexity to the problem. This is the stability of the preparations once they have been produced (expiry dates). Some preparations with a very limited shelf life should not be produced long before the start of the round and should be delivered first. Otherwise, the delivery person could be led to stop their round in progress in order to go back for the urgent delivery, which would be very detrimental for the overall solution.

6.5.2. *Special case: flow shop workshop and a single vehicle*

Consider that the production workshop is a flow shop-type workshop in m machines, that is, all jobs have the same range and must be carried out first on machine M_1, then on machine M_2 and finally on machine M_m. This is a good approximation of the process as the production of a chemotherapy preparation always follows the same path, first with the doctor's visit, then sterilization and preparation and finally control. We are freeing ourselves here from the allocation problems that complicate the problem.

The problem first consists of scheduling a set J of n jobs on the machines. With every job J_j are associated a noted execution time $p_{i,j}$ (duration of J_j on M_i, a desired delivery date noted d_j and a noted delivery site j). The production site is marked 0. We know $tt_{i,j}$, the time to go from the site i at site j ($0 \leq i, j \leq n$). Once the jobs are finished, the problem is to group them in batches to distribute them. Once the batches are defined, the problem is to determine a route to follow to distribute all the jobs in each batch. We note D_j the delivery date of J_j. The delay of J_j is now measured against the delivery date, so we have:

$$T_j = \max (0, D_j - d_j)$$

Several criteria can be defined, for example, the criterion T_{max}, already addressed in section 2.1.1, the sum of the delays denoted by $\sum T_j$ and the number of late deliveries denoted by $\sum U_j$, where $U_j = 1$ if $T_j > 0$, and 0 otherwise.

EXAMPLE.– Consider a problem where scheduling is done on a single machine, with the following six jobs. The distance matrix (7×7) is as follows (note that, in the general case, this matrix is not symmetrical). The locations of the sites are shown in Figure 6.9, and only one vehicle is available for delivery.

j	1	2	3	4	5	6
p_j	8	7	5	2	3	5
d_j	15	17	20	22	25	30

$$tt_{i,j} = \begin{pmatrix} 0 & 3 & 3 & 3 & 4 & 2 & 3 \\ 3 & 0 & 6 & 1 & 7 & 4 & 5 \\ 3 & 6 & 0 & 5 & 3 & 4 & 1 \\ 3 & 1 & 5 & 0 & 7 & 5 & 5 \\ 4 & 7 & 3 & 7 & 0 & 3 & 3 \\ 2 & 4 & 4 & 5 & 3 & 0 & 4 \\ 3 & 5 & 1 & 5 & 3 & 4 & 0 \end{pmatrix}$$

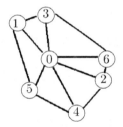

Figure 6.9. *Geographical distribution of sites*

We seek to minimize the sum of delays, $\sum T_j$.

We show in Figure 6.10 two solutions to the problem. In the first solution, the scheduling sequence is $(J_1, J_2, J_3, J_4, J_5, J_6)$; in other words, the jobs are sorted according to EDD (not optimal for the criterion $\sum T_j$ to a machine but optimal for the criterion T_{max}). Batches are $\{J_1\}$, $\{J_2\}$, $\{J_3\}$, $\{J_4, J_5\}$ and $\{J_6\}$. In the $\{J_4, J_5\}$ batch, the delivery sequence is J_5 then J_4. The delivery dates of the jobs are (11, 18, 24, 32, 29, 39), which gives a total delay equal to 28. In the second solution, which does not respect the intuitive order of EDD, the sequence is $(J_4, J_1, J_3, J_5, J_2, J_6)$. Each job forms a batch on its own. Delivery dates are equal to (13, 29, 19, 6, 24, 35), which leads to a total delay of 17, which is much better.

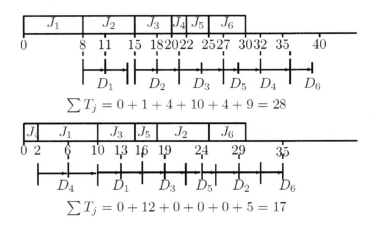

Figure 6.10. *Two examples of scheduling and routing*

Integer linear programming model. A linear integer programming model can be proposed. The first challenge is to link the two levels of planning. The second difficulty consists of finding an effective model for a good resolution by a solver.

Let us consider a flow shop scheduling problem and a single vehicle rounding problem. The following binary decision variables are defined: $z_{j,k}$, $j \in \{1, \dots, n\}$, $k \in \{1, \dots, n\}$ for sequencing the jobs in the workshop; $y_{j,r}$, $j \in \{1, \dots, n\}$, $r \in \{1, \dots, n\}$ for assigning jobs to batches and $x_{i,j,r}$, $i \in \{0, \dots, n\}$, $j \in \{0, \dots, n\}$, $r \in \{1, \dots, n\}$ for job sequencing within batches:

$$z_{j,k} = \begin{cases} 1 \text{ if the work } J_j \text{ is in position } k \\ 0 \text{ otherwise} \end{cases}$$

$$y_{j,r} = \begin{cases} 1 \text{ if the work } J_j \text{ is in the batch/round } r \\ 0 \text{ otherwise} \end{cases}$$

$$x_{i,j,r} = \begin{cases} 1 \text{ if the curve } (i,j) \text{ is in the batch/round } r \\ 0 \text{ otherwise} \end{cases}$$

The following continuous variables are also required: $C_{k,i} \geq 0$, $k \in \{1, \ldots, n\}$, $i \in \{1, \ldots, m\}$ to indicate the end date of the job in position k on the machine M_i, $t_r \geq 0$, $r \in \{1, \ldots, n\}$ the departure date of the round r, $A_j \geq 0$, $j \in \{0, \ldots, n\}$ the time required to deliver the job J_j on its round and $T_j \geq 0$, $j \in \{1, \ldots, n\}$ delay in job J_j.

The objective function is:

$$\text{MIN } \sum_{j=1}^{n} T_j$$

The delay of J_j is greater than or equal to its delivery date $t_r + A_j$ minus its due date d_j, if J_j is on round r. It is expressed by $\forall j \in \{1, \ldots, n\}$, $\forall r \in \{1, \ldots, n\}$:

$$T_j \geq t_r + A_j - d_j - M(1 - y_{j,r}) \qquad [6.10]$$

Each job is only at one position, so $\forall j \in \{1, \ldots, n\}$:

$$\sum_{k=1}^{n} z_{j,k} = 1 \qquad [6.11]$$

In each position, there is only one job, so $\forall k \in \{1, \ldots, n\}$:

$$\sum_{j=1}^{n} z_{j,k} = 1 \qquad [6.12]$$

Each job is necessarily in a batch, so $\forall j \in \{1, \ldots, n\}$:

$$\sum_{r=1}^{n} y_{j,r} = 1 \qquad [6.13]$$

The scheduling part is classic. It reflects the precedence constraints related to the range of jobs and disjunctive resources. The case of working in the first position and the case of the first machine are treated separately in the constraints. The set of constraints is as follows:

$$C_{1,1} = \sum_{j=1}^{n} p_{j,1} z_{j,1} \qquad [6.14]$$

$$C_{k,1} = C_{k-1,1} + \sum_{j=1}^{n} p_{j,1} z_{j,k} \qquad \forall k \in \{2, \ldots, n\} \qquad [6.15]$$

$$C_{1,i} = C_{1,i-1} + \sum_{j=1}^{n} p_{j,i} z_{j,1} \qquad \forall i \in \{2, \ldots, m\} \qquad [6.16]$$

$$C_{k,i} \geq C_{k-1,1} + \sum_{j=1}^{n} p_{j,i} z_{j,k} \quad \forall k \in \{2, \dots, n\}, \quad \forall i \in \{2, \dots, m\} \quad [6.17]$$

$$C_{k,i} \geq C_{k,i-1} + \sum_{j=1}^{n} p_{j,i} z_{j,k} \quad \forall k \in \{2, \dots, n\}, \quad \forall i \in \{2, \dots, n\} \quad [6.18]$$

The part related to routing is as follows. The link between the $x_{i,j,r}$ and $y_{j,r}$ variables is as follows (among others, if J_j is not on round r, then all $x_{i,j,r}$ and $x_{j,i,r}$ variables are equal to 0):

$$\sum_{j=0}^{n} x_{i,j,r} = y_{i,r} \qquad \forall i \in \{1, \dots, n\}, \forall r \in \{1, \dots, n\} \qquad [6.19]$$

$$\sum_{i=0}^{n} x_{i,j,r} = y_{j,r} \qquad \forall j \in \{1, \dots, n\}, \forall r \in \{1, \dots, n\} \qquad [6.20]$$

The following constraints impose that a round cannot begin before the end of the round jobs, nor before the return of the vehicle from the previous round.

$$t_r \geq C_{k,m} - M\big(2 - z_{j,k} - y_{j,r}\big) \quad \forall j \in \{1, \dots, n\}, \quad \forall k \in \{1, \dots, n\}, \forall r \in \{1, \dots, n\} \qquad [6.21]$$

$$t_r \geq t_{r-1} + \sum_{i=0}^{n} \sum_{j=0}^{n} tt_{i,j} \times x_{i,j,r-1} \quad \forall r \in \{2, \dots, n\} \qquad [6.22]$$

The delivery time of J_j in its round is given by $\forall j \in \{0, \dots, n\}$, $\forall j \in \{1, \dots, n\}$, $\forall r \in \{1, \dots, n\}$, $i \neq j$ (with $A_0 = 0$):

$$A_j \geq A_i + tt_{i,j} - M\big(1 - x_{i,j,r}\big) \qquad [6.23]$$

Each round starts from the depot, that is, $\forall r \in \{1, \dots, n\}$:

$$\sum_{i=0}^{n} x_{0,i,r} \leq 1 \qquad [6.24]$$

This model includes $n(n+1)^2 + 2n^2$ binary variables, $0(n^2)$ continuous variables and $O(n^3 + nm)$ constraints, including $2n^2(n+1)$ constraints with a very high constant M.

6.5.3. General case

In the general case, the preparations, once controlled, are taken out of the room through an airlock and placed in a refrigerator awaiting delivery. Many hospital sites have patients waiting for treatment. Several delivery people are responsible for delivering the preparations, usually with one delivery person assigned to a particular site.

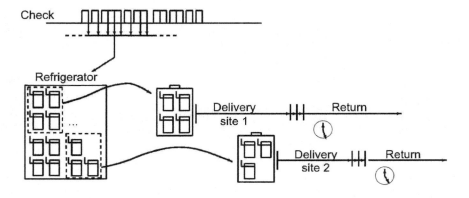

Figure 6.11. *Consideration of distribution*

The solution of the problem for instances of real size as a whole requires the implementation of approached methods. Once again, metaheuristics seem to be very promising methods (Ta *et al.* 2015; Billaut *et al.* 2017).

This problem was at the heart of the ANR ATHENA ANR-13-BS02-0006 project, which dealt more generally with the resolution of complex integrated problems.

6.6. Conclusion

After about 10 years of collaboration between the *Laboratoire d'informatique de l'université de Tours and le CHRU de Tours* (the computer science laboratory of the University of Tours and the CHRU of Tours), in this chapter we come to a synthesis on three problems related to the production of chemotherapy.

The first problem we addressed is the daily planning of chemotherapy production. Two software programs have been developed to solve this problem: one for the planning itself and the other for production traceability. The particularity of the problem lies in the structure of the workshop, where each machine is a mini-workshop composed of a max-batch-type machine on the first level (the sterilizer) and several parallel machines on the second level (the dispensing pharmacists). The whole process ends with a single machine in charge of checking the preparations. The interactive method used to solve the problem was described.

The second problem we addressed concerns the consideration of residues. These are active products used in chemotherapy preparations, which are very expensive and volatile. This brings new complications to the problem: first, the fact that once a preparation is made it must be administered to the patient within a specific time window; second, the desire to reduce the loss of these products implies the introduction of new objective functions, not only linked to the end dates of the jobs, but also to their consumption and the life of the products. Treatment of this problem requires the prior implementation of a system to monitor chemotherapy product stocks. In the case where the production system is reduced to a single machine and for a single product, the problem is close to the two-constraint bin packing problem and we present a linear integer programming model that solves this problem.

The third problem concerns the consideration needed in distribution. Given transport times and production times, this problem is strongly connected to production. In the case where production is carried out in a flow shop, and with only one vehicle for distribution, we present a linear programming model.

Research perspectives on these issues are numerous. The problems are very complex, and the development of efficient methods is a real challenge. The realization of a software suite for an effective implementation of the associated algorithms is also a difficult task, which requires a solid partnership between the various participants of the project.

Another problem related to the production of chemotherapy concerns the scheduling of outpatient appointments to administer treatments. To improve the production process, backlog management and/or distribution planning, patients treated with chemotherapies requiring the same cytotoxic products and/or located in the same department could be scheduled for the same day. To our knowledge, no studies have looked at the coupling of appointment scheduling with chemotherapy production. However, the appointment scheduling problem can itself be very complex when all resource constraints must be taken into account (availability of nurses, doctors, compliance with care protocols, available beds, etc.) (Condotta *et al.* 2014; Hahn-Goldberg 2014).

6.7. Acknowledgments

The authors would like to thank the ANR for funding the project ANR ATHENA ANR-13-BS02-0006.

6.8. References

Alves, C., de Carvalho, J.V., Clautiaux, F., and Rietz, J. (2014). Multidimensional dual-feasible functions and fast lower bounds for the vector packing problem. *European Journal of Operational Research*, 233, 43–63.

Aubert, A., Tournamille, J.F., André, V., De Laguerenne, A., Mazier, A., Billaut, J.C., and Antier, D. (2009). The impact of computer modelling on planning chemotherapy preparations. *Hospital Pharmacy Europe*, (44), 70–71.

Bilgen, B., Ozkarahan, I. (2004). Strategic tactical and operational production-distribution models: a review. *International Journal of Technology Management*, 28(2), 151–171.

Billaut, J.C. (2011). New scheduling problems with perishable raw materials constraints. *16th IEEE International Conference on Emerging Technologies and Factory Automation (ETFA'2011)*, Toulouse, France.

Billaut, J.C. (2014). Deux problématiques abordées au CHRU de Tours. *Journée d'efficience des systèmes de soins* (JESS'2014), École Nationale Supérieure des Mines de Saint-Étienne, France.

Billaut, J.C., Della Croce, F., and Grosso, A. (2015). A single machine scheduling problem with bin packing constraints. *European Journal of Operational Research*, 243, 75–81.

Chen, Z.L. (2004). Integrated production and distribution operations: taxonomy, models, and review. In *Handbook of Quantitative Supply Chain Analysis: Modeling in the E-Business Era*, Simchi-Levi, D., Wu, S.D., and Shen, Z.J. (eds). Kluwer Academic Publishers, Norwell, MA.

Condotta, A., Shakhlevich, N.V. (2014). Scheduling patient appointments via multilevel template: a case study in chemotherapy. *Operations Research for Health Care*, 3(3), 129–144.

Delchaux, D. (2016). *La vérité sur la surenchère des anticancéreux*. [Online]. Available at: https://www.challenges.fr/entreprise/sante/la-verite-sur-la-surenchere-des-anticancereux_16783.

Desrosiers, J., Dumas, Y., Solomon, M.M., and Soumis, F. (1995). Time constrained routing and scheduling. In *Network Routing: Handbooks in Operations Research and Management Science*, Ball, M.O., Magnanti, T.L., Monma, C.L., and Nemhauser G.L. (eds). 8, 35–139, North-Holland, Amsterdam.

Hahn-Goldberg, S., Carter, M.W., Beck, J.C., Trudeau, M., Sousa, P., and Beattie, K. (2014). Dynamic optimization of chemotherapy outpatient scheduling with uncertainty. *Health Care Management Science*, 17(4), 379–392.

Kergosien, Y., Gendreau, M., and Billaut, J.C. (2017). A Benders decomposition-based heuristic for a production and outbound distribution scheduling problem with strict delivery constraints. *European Journal of Operational Research*, 262(1), 287–298.

Maraninchi, D. and Vernant, J.P. (2016). *L'urgence de maîtriser les prix des nouveaux médicaments contre le cancer* [Online]. Available at: sante.lefigaro. fr/actualite72016/03/14/24739-lurgence-maitriser-prix-nouveaux-medicaments-contre-cancer.

Mazier, A., Billaut, J.C., and Tournamille, J.F. (2007). Scheduling activities in a chemotherapy service. *International Conference on Industrial Engineering and Systems Management (IESM'07)*, Beijing, China.

Mazier, A., Billaut, J.C., and Tournamille, J.F. (2010). Scheduling preparation of doses for a chemotherapy service. *Annals of Operations Research*, 178(1), 145–154.

Paillé, Y. (2016). Cancer: la France face au coût des nouveaux traitements et de l'innovation [Online]. Available at: https//www.latribune.fr/entreprises-finance/industrie/chimie-pharmacie/cancer-la-france-face-au-cout-des-nouveaux-traitements-et-de-linnovation-557352.html.

Respaud, R. (2011). Etude de stabilité de médicaments anticancéreux injectables: rapports analytiques et pharmaceutiques. PhD thesis, Université de Tours, France.

T'Kindt, V. and Billaut, J.C. (2006). *Multicriteria Scheduling. Theory, Models and Algorithms*, 2nd edition. Springer-Verlag, Heidelberg, Germany.

Ta, Q.C., Billaut, J.C., and Bouquard, J.L. (2015). Tabu search algorithms to minimize the total tardiness in a flow shop production and outbound distribution scheduling problem. *International Conference on Industrial Engineering and Systems Management (IESM'2015)*, Seville, Spain.

Ullrich, CA. (2013). Integrated machine scheduling and vehicle routing with time windows. *European Journal of Operational Research*, 227(1), 152–165.

Viergutz, C. and Knust, S. (2014). Integrated production and distribution scheduling with lifespan constraints. *Annals of Operations Research*, 213(1), 293–318.

Part 4

Age-appropriate Technologies

Introduction to Part 4

Thus, ever drawn toward far shores uncharted,
Into eternal darkness borne away,
May we not ever on Time's sea, unthwarted,
Cast anchor for a day?
Alphonse De Lamartine

The passing of time, though elusive, leaves a tangible mark on living matter. Aging characterizes an inevitable evolution of the body whose quantification is determined by chronological age. However, this mark of time is highly variable and multifactorial. The chronological age criterion only imperfectly reflects the actual aging of the body, and strong variations can be observed between individuals who are of the same chronological age, which makes the management of aging difficult to achieve.

In Western societies, however, aging continues to occupy an increasing place, disrupting established conceptions in the social, economic and health fields, thus becoming a significant political subject. Health, and therefore medical practice, has been greatly disrupted in the past 50 years (especially due to the appearance of antibiotics) by this change in the age pyramid. From targeted care in a young population, the physician has shifted to multi-dimensional care in an older population. The bronchitis in the young dynamic frame has nothing to do with the bronchitis of the suffering octogenarian. Where doctors will take action within 15 minutes for a young patient with all their bookish knowledge, they should have patience and experience to understand the potential morbid cataclysm of the elderly person's bronchitis.

In negotiating care modalities with the elderly patient, physicians must integrate the physical, psychological, social, cultural and existential dimensions, taking advantage of the knowledge and trust generated by repeated contacts. Indeed, they often know their patient well, and like a cancer specialist, as has already been said before, they live this long devastating agony with a forced fatalism. The older the patient gets, the more they must have a synthetic mind where the part related to the context of their thinking, aging with the changes that accompany it in the patient, increases the complexity of this approach. They can no longer consider the reason for their visit in a single clinical dimension but rather according to all the dimensions provided by the associated co-morbidities whose prevalence increases with age. They must then include this management in a public health approach aimed at anticipating these co-morbidities in a clinical and economic ethical concern. In this perspective and in view of the increasing aging of the population, the daily organization of the doctor becomes increasingly difficult.

With an estimated average consultation time of 16 minutes, quality management assurance forces physicians to use effective related tools to help them maintain their role. These tools can have two vocations: a diagnostic or prognostic role in a pathological state already advanced and a preventive anticipatory role in a healthy state. This last vocation has been explored many times but still opens up many possibilities for reflection.

Comparison of Two Hospitalization Admission Pathways in Geriatrics, Either Directly through a Telephone Line or Hotline or After a Visit to an Emergency Department

7.1. Introduction

Elderly polypathological patients (with several associated diseases) are difficult to manage. They have high rates of medical visits, are more often hospitalized and have a higher rate of emergency room visits than the general population (Salisbury *et al.* 2011; Le Pape *et al.* 1997).

For general practitioners in France, *Services d'accueil d'urgence* or SAU (emergency intake departments) are the fastest way to find a solution to a complex situation (Vedel *et al.* 2009). The ANAP[1] study showed that hospital admissions via emergency departments for people over 80 on average account for 41% of stays and are even the majority in certain regions (Anap 2012; Anap 2011).

Chapter written by Laure MARTINEZ, Marie-Ange BLANCHON, Thomas CELARIER and Marianne SARAZIN.
1 *L'Agence nationale d'appui à la performance des établissements de santé et médico-sociaux* (the National Performance Support Agency for Health and Medico-social Institutions).

Many studies have shown that the passage of elderly patients through emergency departments carries significant risks for the maintenance of autonomy, generating a tendency for bedridden patients as a function of the time spent in this type of department, or promoting the appearance of acute confusion triggered by stress (Wargon *et al.* 2008; Creditor *et al.* 1993; Hoenig *et al.* 1991; Neouze *et al.* 2012; Mazières *et al.* 2011; McCusker *et al.* 1999, McCusker *et al.* 2001). More and more doctors are aware of the harmful effects for their elderly patients of these hospitalizations by passing through the emergency department (Vedel *et al.* 2009).

At the same time, inappropriate use of emergency departments contributes to disorganization of these services. Emergency departments are quickly saturated. The quality of care for patients may be affected, with longer waiting periods, longer stays and difficulties in referring patients from the emergency department to more appropriate departments (Wargon *et al.* 2008; Condelius *et al.* 2008; Noel *et al.* 2007; Laux *et al.* 2008; Richardson 2006). Moreover, this large number of hospitalizations poses an economic problem. Indeed, according to the *Haut Conseil pour l'Avenir de l'Assurance Maladie*, HCAAM (French High Council for the Future of Health Insurance), the additional costs induced by segmentation and inadequate responses to care, such as the unjustified use of hospitalization, have been estimated at 2 billion euros.

To optimize care pathways for elderly patients with multiple diseases, the French government has planned measures via the *loi Hôpital, Patients, Santé et Territoires*, HPST (Hospital, Patients, Health, and Territories law) of July 2009: each territory has been equipped with a geriatric network for the provision of a telephone "hotline" for city doctors. This would allow them direct access to a geriatric physician to respond to their request. This system was designed, among other things, to encourage direct hospitalizations in geriatric wards without going through emergency departments.

Few studies have assessed the real impact of these hotlines and their relevance (Bailly *et al.* 2014).

QUESTION.– Is this of any interest to the patients in terms of prognosis? Does it really avoid emergency department visits? Do elderly patients have an interest in this type of hospitalization?

Given the lack of knowledge on the impact of the hotline system on the care of elderly people, the main objective of this work is to compare the care pathways of patients who could be hospitalized directly in the geriatric ward via a hotline versus those who transited through the emergency department. The starting hypothesis was that patients hospitalized via a hotline had a better prognosis than those hospitalized after an emergency room visit.

This work was carried out by relying on the "hotline" system already existing in the geriatrics department of the Saint-Etienne University Hospital Center.

7.2. Materials and methodology

7.2.1. How the hotline works

The "hotline" is a telephone line dedicated to the geriatrics department, with the aim of allowing direct access to specialized medical advice for doctors practicing outside the hospital. This telephone line is open Monday through Friday from 9 am to 6 pm. It is meant for general practitioners, *SOS médecins* (emergency department physicians), and EHPAD[2] coordinating physicians in the Saint-Etienne city area. Geriatricians give telephone advice on the problem submitted to them. This advice may be therapeutic advice, a request for consultation, a request for immediate or deferred hospitalization, or referral assistance.

Every day, the geriatric department with this hotline reserves three beds until 1:30 pm to respond as quickly as possible to hospitalization requests made via the hotline itself.

It was specified that it was not intended to respond to requests for admission to institutions or to resolve psychological or behavioral problems.

This hotline was set up in March 2013 at the Saint-Etienne University Hospital center.

---

2 *Etablissement d'Hébergement pour Personnes Agées Dépendantes* (Accommodation Establishments for Dependent Elderly People).

7.2.2. *Population studied*

Inclusion of patients occurred between June 1, 2015, and June 1, 2016. The patients included had to be hospitalized in a specialized geriatric ward of the Saint-Etienne University Hospital Center, coming directly from their home.

EHPAD resident patients were not included in the study because of specific care problems nor were patients hospitalized in another specialty department (outside the emergency department) before entering the geriatric department (surgery department, intensive care department).

7.2.3. *Variables studied*

For each patient included, the required information was collected prospectively by the unit's physicians using a questionnaire.

Type of information	Details of variables
Administrative	Surname, first name, date of birth, sex, medical file identification number (IPP[3]), mode of hospitalization: direct following a "hotline" telephone call or after a visit to the emergency department, person having requested hospitalization: doctor on duty, attending physician, family
Social	Living alone or not living alone, presence of caregivers on a daily basis, notion of caregiver exhaustion
Drug treatment	Number of drugs taken and any changes that occurred in the 15 days preceding hospitalization
Patient's condition upon arrival at and discharge from hospital	Current and past illnesses, Charlson score (20-22), ADL (23) and IADL (24) autonomy scales, presence of walking difficulties, level of undernutrition, presence of cognitive disorders with the Mini-Mental State Examination (25), number of hospitalizations during the previous 12 months, GIR[4] test (26)
Orientation of the patient after their stay in the geriatric ward	Referral to a specialized rehabilitation department or patient death

Table 7.1. *The items concerned*

---

3 *Indemnité permanente partielle* (Partial Permanent Allowance).
4 *Groupe iso-ressources* (Iso-resource Group).

7.2.4. *Comparison of the two types of pathways*

Two groups of patients were formed according to the mode of admission into geriatric department:

– the patients entered via a telephone call through the hotline composed the first group (hotline group);

– emergency department patients were the second group.

The judging criteria used to compare the two types of pathways were:

– the length of stay following hospitalization including days in emergency departments for patients hospitalized by this mode;

– the evolution of the loss of autonomy during hospitalization according to two modalities: maintenance of autonomy in relation to the previous state versus degradation of autonomy;

– the release: return home, death, additional specialized re-education, orientation in a medico-social structure; or long stay in a retirement home.

Patients were classified according to the seven main reasons for hospitalization: fall or malaise, confusion, loss of autonomy, general impairment, pain, dyspnea or other reasons.

7.2.5. *Statistical analysis*

The two pathways were compared based on the duration of hospitalization, the delay until the next re-hospitalization, and the status at discharge by varied bi-analysis with chi-squared tests for the qualitative variables and the Student's-test for the quantitative variables and a risk of error considered at 5%.

7.3. Study results

A total of 520 patients were included in the study.

7.3.1. *Socio-demographic characteristics and care pathways of the population studied*

General population characteristics are summarized in Table 7.2.

	Hotline n=175	Emergency Dept. n= 345	p
Age (avg +/– SD)	86.1 +/–0.9	86.3 + /–0.6	0.69
Woman (%)	67.4 +/–6.9	65.8 +/–5	0.78
Isolation Lives as couple (%) Lives alone (with children) (%) Lives alone (without children) (%)	20 42.9 37.1	18 43.8 38.3	0.85 0.85 0.85
Home Helpers (%)	34.1	30.3	0.62
Charlson score (avg +/- SD)	6.4 +/–0.3	6.5+/–0.2	0.57
MMSE/30 (avg +/- SD)	21+/–0.8	20+/–0.8	0.09
ADL/6 (avg +/- SD)	4.3+/–0.2	4.3+/–0.2	0.96
IADL/4 (avg +/- SD)	1.6+/–0.2	1.5+/–0.1	0.48
Number of usual treatments > 4/day (%)	73.7	71.3	0.77
Number of hospitalizations in the previous 12 months (avg + /- SD)	1.45+/–0.1	1.32+/–0.07	0.04

Table 7.2. *Socio-demographic and geriatric characteristics of the population of the 520 patients hospitalized for a short geriatric stay either via the hotline or via the UAS.*

The average age is 86.2 years with the majority being women: 345 women (66.3%) versus 175 men (33.7%).

A total of 175 patients were directly admitted to the geriatric department via the hotline, that is, 33%, and 345 patients went through the emergency department, which represents 66% of the patients included. The socio-demographic characteristics of patients hospitalized via the hotline were not significantly different from those of patients hospitalized via the emergency department.

The two groups are comparable in terms of patient status with a Charlson score of 6+/–0.3 on average in both groups, an MMSE[5] of 21+/–0.8 in hotline patients and 20+/–0.8 in patients hospitalized via emergency departments. 73.7% of patients admitted via the hotline consumed more than four drugs daily compared to 71.3% for those hospitalized via the emergency room (p=0.77). On the other hand, the 40.8% of patients hospitalized via the hotline who changed treatment in the 15 days preceding hospitalization are compared to only 15.6% (p<0.001) for patients hospitalized via emergency departments.

The three main reasons for hospitalization via the hotline are alteration of general condition (17.7% of reasons), pain (15.4% of reasons) and loss of autonomy (6.3% of reasons), while, in the group hospitalized via emergency departments, the most common reasons for hospitalization are falls and discomfort (38.8%) and dyspnea (14.8%).

7.3.2. Comparison of pathways according to the two hospitalization admission modes

As described in Table 7.3, the length of stay is significantly shorter for hospitalization via the hotline, 11.6 +/–0.8 days on average for direct hospitalization against 14.1+/–0.6 days for an emergency visit (p<0.05), that is, 28% shorter. There is a slight disparity with lengths of stay ranging from 10.8 days to 12.3 days for direct hospitalization and from 13.5 days to 14.7 days for patients who went through emergency departments.

Patients who were admitted to the emergency department for a short stay were re-hospitalized more quickly. Among the 170 patients readmitted to the hospital, the average duration before readmission was 29.5+/–5.9 days for patients hospitalized via the hotline, while those admitted via emergency departments were hospitalized within 24.1+/–3.7 days (p<0.05).

---

5 Mini-Mental State Examination.

The study shows a better prognosis at immediate discharge from hospital with more returning home and fewer deaths after direct admission, but this has not been shown significantly. In fact, 61.7% of the patients who were directly admitted returned home, while only 55.1% of those admitted via emergency departments returned home (p=0.46). A higher number of deaths should be noted after a visit to the emergency room (29.6% vs. 27.9%, p<0.46). The number of admissions to a medico-social structure after the geriatric stay was equal in both groups (8% vs. 7.6%). Of the hospitalized patients, 5.7% went to the SSR[6] compared to 7.6% who went through the emergency room.

	Hotline n=175	Emergency Dept. n=345	p
Length of stay following hospitalization (average +/– SD) in days	11.6 +/–0.7	14.1+/–0.6	<0.001
Output mode Return home (%)	61.7	55.1	0.46
Deaths (%)	24.6	29.6	0.46
Additional Special Rehabilitation (%)	5.7	7.6	0.46
Orientation in a medico-social structure or long stay (%)	8	7.6	0.46
For patients who were re-hospitalized Average time at home before needing for hospital services again (day +/–SD)	29.5 +/–5.9	24.1+/–3.7	0.04

Table 7.3. *Comparison of the length of stay, discharge modes and re-admissions of 175 patients hospitalized via the hotline versus the 345 hospitalized after passing through the emergency department.*

Moreover, a better level of autonomy is observed at the end of hospitalization: 73.8% of patients leave without loss of autonomy after having been hospitalized directly via the hotline versus 67.7% after having passed through the emergency department p<0.0001.

---

6 Soins de suite et de réadaptation (Follow-up Care and Rehabilitation).

Finally, the comparison of hospitalization durations between the two groups shows a significantly shorter average hospitalization duration for each of the reasons for hospitalization in the hotline group. The results are presented in Table 7.4.

Reason for hospitalization: Total number (%) Average length of stay in days (average+/–DS)	Hotline n=175	Emergency Dept. n=345	p
Fall or discomfort (%)	36 (20.5) 11.2+/–1.1	134 (38.8) 14.9+/–0.9	<0.0001
Confusion or behavioral disorder n (%)	17 (9.7) 13.5+/–2.8	26 (7.5) 14.9+/–1.8	<0.0001
Loss of autonomy n (%)	11 (6.2) 11+/–3.5	12 (3.4) 16+/–3.7	<0.0001
AEG[7] n (%)	32 (18.2) 11.8+/–1.5	21 (6.0) 13.5+/–1.8	<0.0001
Pain n (%)	27 (15.4) 12.7+/–2.8	15 (4.3) 13.3+/–2.9	<0.0001
Dyspnea n (%)	12 (6.8) 8.4+/–2.5	51 (14.7) 13.6+/–1.3	<0.0001
Other n (%)	39 (22.2) 11.6+/–1.1	85 (24) 14.3+/–1.7	<0.0001

Table 7.4. *Comparison of reasons for hospitalization according to the two pathways.*

7.4. Discussion

This work has highlighted the value of using a hotline in the organization of a department specialized in geriatric patient care.

Indeed, hospital stays are significantly shorter when patients enter the ward directly through this system than when they return after a visit to the emergency department, without a factor specific to the patient being able to explain this difference. At the same time, it was noted that the number of hospitalizations of elderly people via the emergency department of the Saint-Etienne University Hospital Center has been falling since 2013, when the hotline was set up. These two analyses complement and corroborate the

7 Altération de l'état général (Alteration of the General State of Health).

results of another study that showed a decrease in the number of hospitalizations via an emergency department for an acute problem in elderly patients living alone at home since the setting up of a hotline in the geriatric ward of the hospital concerned (Bailly *et al*. 2014).

The increase in the number of beds in geriatric department, an action carried out by certain hospital centers, has so far not been sufficient to significantly reduce the rates of hospitalization in these departments following a visit to the emergency department (Defèbvre *et al*. 2016). The admissions made directly remain, in spite of this increase in the difficulty of organizing beds. The development of hotlines should facilitate this organization, as this work has shown.

Concerning the length of hospitalization for any course of care, the value obtained, 13 days, at the Saint-Etienne University Hospital Center does not differ from the national average values (Defèbvre *et al*. 2016). These durations included the time spent in emergency departments for the "emergency admission" branch. They are significantly longer by three days on average than the durations of stay for patients who entered directly into the geriatric department. This reinforces the interest of direct admission for an elderly person in a geriatric department. The benefit of the emergency department could be a rapid assessment of the patient's state of health. However, extending the length of the stay when the patient goes through the emergency department does not bring any element in favor of this benefit.

Previous work has shown that direct admission rates to geriatric departments are between 20 and 25% of admissions on average, and transfer rates from an emergency department represent 60% of admissions. In other studies, it was found that the main reasons for hospitalization for patients entering geriatrics directly were altered general condition and loss of autonomy, while, for patients entering after an emergency department visit, the reasons were falls, discomfort or respiratory problems (Neouze *et al*. 2012). This work confirms all these results (Defèbvre *et al*. 2016).

There is no statistically significant difference between the two types of management for the place following hospitalization of patients. However, there is a tendency for more people to return home when admitted directly to a geriatric ward (almost 62% vs. only 55%) than when the patient goes through an emergency department. This rate is higher than that observed in hospitals in Nord Pas de Calais, where the average number of patients returning home after direct hospitalization in the geriatric ward was 56% (Defèbvre *et al.* 2016).

Re-admissions following this stay in the geriatric ward occur on average later and less frequently after direct hospitalization in the geriatric ward than when there was a visit to the emergency department. Among the studies that have attempted to identify risk factors likely to explain these re-admissions, a study carried out at the Bordeaux University Hospital Center showed that it was important to maintain a certain length of stay to avoid a new recourse to the hospital service (Pérès *et al.* 2002). The duration envisaged is approximately 10–11 days, which corresponds to that observed in this work for a mode of hospitalization by direct admission.

Finally, other works (Bertrand 2006; Bas 2007; Fried *et al.* 2004; Bellou *et al.* 2003) showed that a geriatric admission system directly coupled with a dynamic network of care between professionals working outside the hospital and hospital professionals improved patients' functional prognosis. This work does not conclusively show a better functional prognosis after direct admission to geriatrics, but it does suggest that a higher proportion of patients admitted directly retain their autonomy and that loss of autonomy is more frequent after a visit to the emergency department.

7.4.1. *Limits*

The reduction in the length of stay generates a reduction in the costs of care as well as the exclusion of emergency departments from these pathways. However, although a saving is made, the organization of a geriatric department, including a hotline, induces costs that this work could not unfortunately identify and calculate, in particular: the increased cost generated by the time spent by geriatricians on this telephone line and the increased cost of beds waiting to be filled.

If it is worthwhile to think that it is useful for elderly people to avoid complex hospital pathways by avoiding emergency departments, it is then relevant to know the assessment and opinion of their doctor. Their doctor may be able to advise on accessibility needs or any adaptations required for patients who take this direct route and avoid the emergency department (extension of hotline timetables, teleconsulations, etc.).

Finally, this work was carried out over only one year, which made it possible to work on a substantial sample which no other work has so far carried out. On the other hand, it could only be done on one establishment, which makes the reproducibility of this organization more uncertain for other French hospitals and towns whose organization of the care route would be different.

7.4.2. Additional work in progress

Given the frequency of hospitalizations of elderly patients with several concomitant illnesses, it is important to generalize hospitalization pathways that make it possible to avoid emergency department visits that generate stress and an increase in dependence. The possibility of a direct telephone call from a town doctor to a geriatrician seems relevant to facilitate the short circuit to the emergency department, the coherent organization of appropriate care and the reduction of time spent in hospital. This device should be used more by general practitioners.

This work has provided arguments in favor of the role of the hotline in shortening the length of stay. However, we do not exclude the possibility that the clinical state of the patients and their social situation may explain a medical behavior still very dependent on the underlying state of the patient and not on the tools made available to integrate the patient into a hospital process. It would therefore be interesting to complete it by analyzing in depth the reasons or reluctance of doctors to use the hotline. A simulation study is underway to try to adapt bed management in geriatric departments to this new hospitalization mode, taking into account the different design possibilities.

7.5. Conclusion

Given the frequency of hospitalizations of elderly polypathological patients, it is important to organize short pathways to reduce the incidence of dependence caused by inappropriate passage via the emergency department. The possibility of a direct telephone call from a geriatrician's city doctor enables better care choices, which seems to shorten the overall time spent in hospital if the data from this work are considered. This proposal seems to be interesting and deserves to be evaluated in other geriatric fields before its widespread use.

> **KEY POINTS**
>
> – Elderly polypathological patients are more frequently hospitalized than the rest of the general population and these hospitalizations are more frequently made via emergency departments.
>
> – Stays that do not include a visit to emergency departments are harmful to maintaining patient autonomy; hence, it is important to organize short pathways to reduce the incidence of dependence.
>
> – The implementation of a direct telephone line for city doctors to a geriatrician seems to shorten the time spent in hospital.
>
> – This system must be evaluated in other geriatric fields before being implemented.

7.6. References

ANAP. (2011). Les parcours de personnes âgées sur un territoire: retours d'expériences. Report, Agence nationale d'appui à la performance des établissements de santé et médico-sociaux, Paris.

ANAP. (2012). Les parcours de santé des personnes âgées sur un territoire : réaliser un diagnostic et définir une feuille de route pour un territoire. Report, Agence Nationale d'appui à la performance des établissements de santé et médico-sociaux, Paris.

Bailly, A. (2014). Intérêt d'une ligne téléphonique directe (Hotline) destinée aux médecins pour limiter le passage aux urgences des personnes âgées analyse de 198 appels. PhD thesis, Université Jean Monnet, Saint-Étienne.

Bellou, A., de Korwin, J.-D., Bouget, J., Carpentier, F., Ledoray, V., Kopferschmitt, J., and Lamberts, H. (2003). Place des services d'urgences dans la régulation des hospitalisations publiques. *La revue de médecine interne*, 24(9), 602–612.

Jeandel, C., Pfitzenmeyer, P., and Vigouroux, P. (2006). Un programme pour la gériatrie: 5 objectifs, 20 recommandations, 45 mesures pour atténuer l'impact du choc démographique gériatrique sur le fonctionnement des hôpitaux dans les 15 ans à venir. Report, French Ministère de la santé et des solidarités, Paris.

Condelius, A., Edberg, A-K, Jakobsson, U., and Hallberg, I.R. (2008). Hospital admissions among people 65+ related to multimorbidity, municipal and outpatient care. *Arch Gerontol Geriatr.*, 46(1), 41–55.

Creditor, M.C. (1993). Hazards of hospitalization of the elderly. *Ann Intern Med.*, 118(3), 219–23.

Defèbvre, M.-M. and Puisieux, F. (2016). Les filières gériatriques en Nord-Pas-de-Calais : analyse huit ans après la circulaire de 2007. *Revue d'Épidémiologie et de Santé Publique*, 64(5), 341–349.

Fried, L.P., Ferrucci, L., Darer, J., Williamson, J.D., and Anderson, G. (2004). Untangling the concepts of disability, frailty, and comorbidity: implications for improved targeting and care. *J Gerontol A Biol Sci Med Sci.*, 59(3), 255–263.

Hoenig, H.M., and Rubenstein, L.Z. (1991). Hospital-associated deconditioning and dysfunction. *J Am Geriatr Soc.*, 39(2), 220–222.

Laux, G., Kuehlein, T., Rosemann, T., and Szecsenyi, J. (2008). Co- and multimorbidity patterns in primary care based on episodes of care: results from the German CONTENT project. *BMC Health Serv Res.*, 8(1), 14.

Le Pape, A., and Sermet, C. (1997). La polypathologie des personnes âgées, quelle prise en charge à domicile? Presentation, INSEE-CREDES "Soigner à domicile" Congres, Paris, 11–12 October.

Mazière, S., Lanièce, I., Hadri, N., Bioteau, C., Millet, C., Couturier, P., Gavazzi, P. (2011). Predictors of functional decline of older persons after an hospitalisation in an acute care for elder unit: importance of recent functional evolution. *Presse Médicale*, 40(2), e101–110.

McCusker, J., Bellavance, F., Cardin, S., Trépanier, S., Verdon, J., and Ardman O. (1999). Detection of older people at increased risk of adverse health outcomes after an emergency visit: the ISAR screening tool. *J Am Geriatr Soc.*, 47(10), 1229–1237.

McCusker, J., Verdon, J., Tousignant, P., de Courval, L.P., Dendukuri, N., and Belzile, E. (2001). Rapid emergency department intervention for older people reduces risk of functional decline: results of a multicenter randomized trial. *J Am Geriatr Soc*, 49(10), 1272–1281.

Neouze, A., Dechartres, A., Legrain, S., Raynaud-Simon, A., Gaubert-Dahan, M.-L., and Bonnet-Zamponi, D. (2012). Hospitalization of elderly in an acute-care geriatric department. *Geriatr Psychol Neuropsychiatr Vieil*, 10(2), 143–150.

Pérès, K., Rainfray, M., Perrié, N., Emeriau, J.P., Chêne, G., and Barberger-Gateau, P. (2002). Incidence, risk factors and adequacy of early readmission among the elderly. *Rev Epidemiol Santé Publique*, 50(2), 109–119.

Richardson, D.B. (2006). Increase in patient mortality at 10 days associated with emergency department overcrowding. *Med J Aust*, 184(5), 213–216.

Salisbury, C., Johnson, L., Purdy, S., Valderas, J.M., and Montgomery, A.A. (2011). Epidemiology and impact of multimorbidity in primary care: a retrospective cohort study. *Br J Gen Pract J R Coll Gen Pract.*, 61(582), e12–21.

Vedel, I., De Stampa, M., Bergman, H., Ankri, J., Cassou, B., Blanchard, F., and Lapointe, L. (2009). Healthcare professionals and managers' participation in developing an intervention: a pre-intervention study in the elderly care context. *Implement Sci IS.*, 4, 21.

Wargon, M., Coffre, T., and Hoang, P. (2008). Durée d'attente des personnes agées aux urgences et dans l'unité d'hospitalisation des urgences. *Journal Européen des Urgences*, 17(HS 1), 118–120.

8

Therapeutic Education for the Patient over 75 Years Old Living at Home

8.1. Introduction

A fall, a frequent problem for elderly patients, corresponds to the action of falling to the ground against their will (Beauchet *et al.* 2011). Of multifactorial origin, it is the main cause of accidental death (Dargent Molina *et al.* 1995) and is identified as a powerful predictor of institutionalization for the elderly (Gillepsie *et al.* 2012). The Haute autorité de santé (HAS) has defined repeated falls as the occurrence of at least two falls in a time interval ranging from 6 to 12 months. (Haute autorité de santé 2009). It considers falling as an indicator of poor health status and a marker of frailty in the elderly (Avreux *et al.* 2002). According to the Institut national de veille sanitaire, INVS, (National institute of elderly health) (Institut de veille sanitaire 2013), every year 450,000 people over the age of 65 fall and 9,300 die. More than half of the falls occur at home. According to recent estimates, one in two people over the age of 80 fall at least once a year. Falls should be considered and managed like any other chronic disease in the elderly. Its prevention is a major public health issue. The multidisciplinary and overall educational approach in the elderly patient with a high risk of falling was cited as an effective therapeutic possibility (Legrain *et al.* 2014; D'Ivernois *et al.* 2013; Pin *et al.* 2015; Bamgbade *et al.* 2016; Robinet *et al.* 2014; Cornillon *et al.* 2002; Puisieux *et al.* 2014).

Chapter written by Justine DIJON, Marianne SARAZIN, Vincent AUGUSTO, Thomas FRANCK and Régis GONTHIER.

Section 84 of the HPST[1] Act (Loi 2009) has included therapeutic patient education in the management of chronic diseases. These pathologies, by being inscribed in the duration, are likely to involve complications and repercussions including an increase in mortality and an alteration of the quality of life (Simon *et al.* 2013).

Therapeutic education is an innovative method involving the patient and caregivers in patient management (D'Ivernois and Gagnayre 2013; Pin *et al.* 2015; Bamgbade and Dearmon 2016; Robinet and Puisieux 2014; Cornillon *et al.* 2002; Puisieux *et al.* 2014; Loi 2009; Simon *et al.* 2013; D'Ivernois and Gagnayre 2011a). This management has already proven its effectiveness in the management of chronic diseases, such as diabetes and heart failure (Legrain *et al.* 2014; D'Ivernois *et al.* 2013; Roussel 2015; Dali *et al.* 2016), to improve patients from a medical perspective by reducing readmissions and improving the quality of life. Therapeutic education provides patients with a privileged forum to better understand their problem, their expectations, and thus improves their care (D'Ivernois and Gagnayre 2013; Mauduit 2014; D'Ivernois and Gagnayre 2011b, Sandrin-Berthon 2009 ; Lagger *et al.* 2008).

This method would seem to have a major interest in helping older people understand the links between falling and their multiple consequences (physical, environmental and psychosocial) while adapting to the wishes, skills and problems presented by each person (Simon *et al.* 2013; Sandrin-Berthon 2009; Lagger *et al.* 2008). The method would assess self-care skills and coping skills (D'Ivernois *et al.* 2011a) of the patient, their diseases like the occurrence of falls, often considered a source of chronicity, loss of autonomy and institutionalization (Puisieux *et al.* 2014; Lagardère *et al.* 2013; Hill *et al.* 2017; Legrain 2014). Indeed, following a fall, elderly patients not only limit their activity to avoid the physical consequences, but also restrict their social life because of apprehension and persistent fear during travel, a real psychomotor maladjustment syndrome (Kendrick *et al.* 2014). Home support may then be compromised if falls occur multiple times.

In 2011, a day-only hospital service called *Hôpital de jour* (HDJ) was created at the *Centre hospitalier et universitaire de Saint-Etienne*, which has developed a multidisciplinary therapeutic education program validated by the Agence régionale de santé (ARS) (French regional health agency), for people over 75 years of age who have fallen while living at home. Patients participating

---

1 Hôpital, patients, santé et territoire (HPST) (French hospital, patients, health, territories act).

in this program benefit from a standardized geriatric assessment and a shared educational assessment in order to target their own needs. The aim of this program has been to offer specific, multidisciplinary and personalized care to elderly patients, to avoid hospitalization in a typical geriatric ward, to limit the physical, psychological and social consequences of the fall and to maintain them as well as possible at home with a good level of autonomy.

The objective of this work was to evaluate the fate of these patients in day hospitals at three and six months from the beginning of treatment, to recognize the impact of the occurrence of new falls, the perceived quality of life, and the fear of falling compared to a population that did not benefit from therapeutic education.

8.2. Materials and methodology

8.2.1. *Study type*

This is an observational, prospective, descriptive, non-randomized, monocentric cohort study (by the *Centre hospitalier universitaire de Saint-Etienne*), following two parallel unmatched groups: a group participating in therapeutic education in the day hospital and a group not participating in therapeutic education, recruited from other geriatric departments.

8.2.2. *Population studied*

8.2.2.1. *Recruitment and inclusion criteria*

The inclusion of persons in the "therapeutic education" group was carried out in the chronological order of patient arrivals in day hospitals from home and on the initiative of a doctor following a fall. The group not involved in therapeutic education was formed from patients hospitalized in a geriatric ward following a fall. The selection was made over a one-year period between 2016 and 2017. Patients were selected without any preconditions other than a refusal to participate in the study and the presence of criteria described in the following section.

Criteria for the inclusion of selected groups			
Age ≥ 75 years	MMSE ≥ 20/30	At least one fall over the course of 6 months	Living at home

8.2.2.2. *Description of the multidisciplinary therapeutic education program set up at the day hospital*

The primary objectives of the program were to evolve the usual care of the elderly patient who had fallen towards a therapeutic education process by promoting their adherence to the program, by actively involving them in the management of their problem as well as their caregivers, despite very varied associated pathologies.

How patients are recruited for the day hospital				
Medical consultation with a geriatrician from the clinical gerontology center of the *Centre hospitalier universitaire de Saint-Etienne*	At the request of a doctor from a geriatrics department of the geriatrics center of the *Centre hospitalier universitare de Saint-Etienne*	By the general practitioner following repetitive falls	By a specialist doctor (ENT specialist, neurologist, rheumatologist)	Saint-Etienne home care network

The workshops were designed using an editorial protocol by staff trained in conjunction with management and responsible physicians (Legifrance 2015; Rey *et al.* 2016). The choice of workshops was defined at the patient's entrance and during the shared educational assessment. Some workshops were systematic for all patients (balance and walking work, floor reading if possible in physiotherapy and occupational therapy, and a home risk workshop); the other workshops were offered according to the patient's problems. Patients had to come twice a week for 12 weeks.

– **Psychological:** *Group management, presence of the psychologist*	– Dealing with the fear of falling – Photolanguage workshop
– **Aesthetic:** *Individual care, IDE nurse and trained AS caregiver*	– Body image enhancement workshop

– **Work on balance and walking:** *Individual and group care, occupational therapist, physiotherapist, physical education instructor, podiatrist*	– Learning about the ground survey – Double task (setting up technical assistance) – Exercise training – Muscle strengthening – Proprioceptive work – Vestibular work – Coordination – Foot care, footwear
– **Work on risks at home:** *Individual and collective care* *Occupational therapist, IDE nurse, care assistant*	– Role-playing with the "outdoor outings" workshop – Transfers – Redesign of the home – "Risks at home" group workshop (IDE nurse, care assistant, occupational therapist)
– **Nutrition:** *Individual and collective care* *Dietician, IDE nurse, healthcare aide*	– Fall and nutrition workshop – Dietary maintenance
– **Medication:** *Individual and collective care, pharmacist, doctors, IDE nurse*	– Fall and medication workshop – Individual interview

Table 8.1. *Description of the different therapeutic workshops*

8.2.3. *The study process*

8.2.3.1. *Collecting patient information*

The information was collected by the doctors treating patients in day hospitals or geriatric departments according to a questionnaire defined when it was included in the study (time 0 or T0), after three months (time 1 or T3) and after six months (time 2 or T6).

8.2.3.2. *Study follow-up*

For the "therapeutic education" group

Patients who benefited from therapeutic education were followed regularly at the day hospital. A "fall" calendar was given to them upon entrance into the study. A weekly multidisciplinary meeting made it possible to follow the evolution of the patients. Any difficulties were immediately reported to the study's principal investigator. The end of treatment in therapeutic education occurred on average within 12 weeks after entry and was marked by a medical visit in which time for evaluation and final synthesis was included. A summary letter was sent to the attending physician informing them of the treatment and important medical and biological data. The set of data collected at the end of the treatment allowed them to define the second time of the study: the time corresponding to the end of the treatment in therapeutic education and the discharge from the day hospital. The third phase of the study took place 3 months after discharge from the day hospital via a telephone interview. This made it possible to retrieve data concerning the main judgment criterion (occurrence of a fall during the last 3 months) and the two secondary criteria (evaluation of the fear of falling and evaluation of the elderly person's quality of life by the EQVPA (life quality scale for the elderly) scale.

For the Hospitalized group

Patients were hospitalized in the geriatric ward. At the end of their stay, they all returned home without additional hospital care. At 3 months and 6 months from the start of their care, a telephone interview was conducted to obtain the data concerning the main judgment criterion (occurrence of a fall during the last 3 months) and the two secondary criteria (evaluation of fear of falling and evaluation of quality of life by the EQVPA scale).

8.2.3.3. *End of the study and flow chart*

The end of the study was effective after the third phase for all patients. The duration of participation in the study for one patient was 6 months.

Lost follow-ups

Patients considered lost and in need of a follow-up were those who could not be reached by telephone. Patients who no longer wished to participate in the study, who were again hospitalized and those who returned to EHPAD were considered as therapeutic reorientations. Those entering institutionalization left the study because the inclusion criteria included living at home.

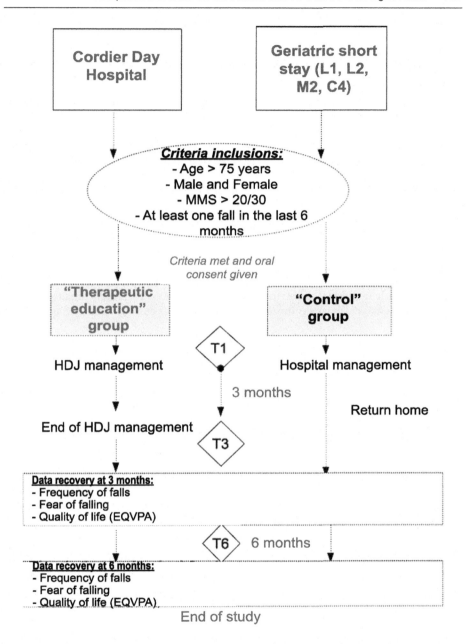

Figure 8.1. *Flow chart. For a color version of the figure,
please see www.iste.co.uk/sarazin/health.zip*

8.2.4. *Data collected*

A summary sheet containing the main data was completed at all three study times for both populations.

---

1) Socio-demographic criteria: age, sex, place of residence, education level classified according to the 6 INSEE training levels (30) and cognitive status evaluated by the MMSE score (/30) at the patient's entry.

2) Clinical and paraclinical characteristics: assessment of autonomy using the ADL (/6) (activities of daily living) and IADL (instrumental activities of daily living) (/4) scales at the patient's entrance, the number of drugs present on the prescription at the patient's entrance, polymedication defined by a number of drugs greater than 4 at the patient's entry (/31), the Charlson score calculated at the patient's entry (/32), the technical aids present at the patient's entry, the patient's entry weight in kg, the albuminemia corrected for CRP at the patient's entry in mg/L.

3) Study endpoints: number of falls reported by the patient, fear of falling measured on a numerical scale from 0 to 4 (no fear to extremely intense fear), validated quality of life scale for the elderly (EQVPA) over 65 years of age (/15) (33).

8.2.5. *Economic evaluation*

The evaluation was carried out to determine the possible cost and revenue of each pathway. For full hospitalization, expenses were based on the value of the average cost of a short-stay bed including staff and operating costs at the *Centre hospitalier universitaire de Saint-Etienne* and the revenue from the stay according to the principle of activity-based pricing for each hospitalized patient. For day hospitalization, the sum of all personnel costs and operating costs was calculated on the basis of those of the day hospital. No additional costs were taken into account (transport, cost of complications, etc.). Revenues were based on the amount owed by patients to the day hospital, established as a lump sum for each day. Personnel costs are based on the public hospital's salary scales, considering the status of each employee.

8.2.6. *Statistical tests*

The quantitative variables were presented in this work with means, standard deviations (m ± sd) and the 95% confidence interval ([95% CI]). The percentages and numbers described the qualitative variables. A chi-squared test compared the qualitative values. The quantitative variables studied were described and compared for each of the two groups using the Mann–Whitney non-parametric test with a $p < 0.05$ value to retain significance. The statistics were conducted using the R software.

The primary endpoint was the number of falls reported at follow-up by patients. Secondary judgment criteria were fear of falling (/4) and the quality-of-life scale of the senior rated by the EQVPA (/15).

Revenues and expenses were evaluated on the basis of the 20 patients hospitalized versus 20 patients taken in day hospitalizations knowing that the fixed price is identical from one patient to another. For complete hospitalization, an average of the income from stays was made from the value of the tariff generated for to each stay profile, in turn according to the principle of the tariff according to each hospitalized patient's activity. The corresponding expenditure was calculated on the basis of an average annual cost of a bed costing €70,000 and reduced to a daily average. This number was then multiplied by the number of days in hospital for the 20 patients. For day hospitalization, the total cost was envisaged over three months corresponding to a usual care program; the receipts correspond to the daily fixed price of €636 paid by the patient for each visit, that is, 24 per patient over three months.

8.2.7. *Ethical considerations*

Patients from the therapeutic education or "ETP" group were given a reflection period following admission to the day hospital to give their consent to participate in the therapeutic education program and study. Patients in the Hospitalized group were offered participation in the study when they entered the geriatric short stay and were given a reflection period to give their consent to participate in the study.

A favorable opinion was obtained from the ethics committee of the *Centre hospitalier universitaire de Saint-Etienne* registered under the following reference – IRBN042016/CHUSTE. Each patient received an information leaflet explaining the study. The declaration of the *Commission nationale de l'informatique et des libertés* (CNIL) (French national commission for informatics and liberties) concerning the conservation of biomedical data has been made and a favorable opinion has been obtained.

8.3. Results

8.3.1. *Patient characteristics*

Socio-demographic characteristics: between September 2015 and September 2016, 48 patients were included, including 28 patients in the ETP group and 20 patients in the Hospitalized group. The two groups were comparable on all inclusion criteria. They were mainly women in both groups (82% in the ETP group and 65% in the Hospitalized group). The number of males was higher in the Hospitalized group, but there was no statistically significant difference on the gender criterion ($p = 0.18$). All patients lived at home. On the cognitive level, the mean MMSE score was comparable with an MMSE of 25.5 +/–3.17 in the ETP group versus 25.0 +/–2.92 in the Hospitalized group ($p = 0.57$); for a similar level of study between the two groups with an average of 5.4 +/–0.74 in the ETP group and 5.4 +/–0.93 in the Hospitalized group ($p = 0.92$).

	ETP group n = 28	Hospitalized group n = 20	p
Gender, n (%) [CI 95%]			ns
Women	23 (82) [0.68–0.96]	13 (65) [0.44–0.85]	
Men	5 (18) [0.038–0.32]	7 (35) [0.14–0.56]	
Age, m ± sd [CI 95%]	84.7 ± 4.31 [84.5–84.9]	85.1 ± 5.02 [84.8–85.2]	ns
MMSE score (/30), m±sd [CI 95%]	25.5 ± 3.17 [24.8–25.7]	25.0 ± 2.92 [24.7–25.3]	ns
Study level (Insee score), m±sd [CI 95%]	5.4 ± 0.74 [5.1–5.7]	5.4 ± 0.93 [4.9–5.8]	ns
Place of residence			
At home, n (%)	28 (100)	20 (100)	

ns: not significant.

Table 8.2. *Socio-demographic characteristics of patients in both groups*

Clinical and paraclinical characteristics of patients: these two populations had very high co-morbidity scores. The Charlson score was comparable in both groups with an average of 7.3 ± 1.21 in the ETP group versus 7.1 ± 1.86 in the Hospitalized group (p = 0.33). Autonomy for basic activities of daily living (ADL) was maintained as statistically comparable in both groups with an average score of 5.4 ± 0.72 in the ETP group versus 5.2 ± 0.95 in the Hospitalized group (p = 0.73). Instrumental activities of daily living (IADL) were statistically significantly more disturbed in the Hospitalized group with an average score of 2.2 ± 1.28 versus 3.0 ± 0.87 in the ETP group (p = 0.025). Polymedication was found among 23 patients in the ETP group (82%) and 16 patients in the Hospitalized group (80%); 28.6% of patients in the ETP group (8/28) and 15% in the Hospitalized group (3/20) were patients with more than 10 drugs at baseline. In nutritional terms, the albuminemia corrected by the CRP was statistically higher in the ETP group with an average of 40.4 ± 3.78 g/L compared to 33.6 ± 5.95 in the Hospitalized group for comparable weights (p = 0.000006). Undernutrition defined by albuminemia of less than 35 g/L was found in eight patients in the Hospitalized group (40%) and one patient in the ETP group (3.6%).

In terms of mobility, 100% of patients in both groups had fallen at least once in the past 6 months. Technical assistance by a single cane was found among 8 patients in the Hospitalized group (40%) and 12 patients in the ETP group (42.9%).

	ETP group n = 28	Hospitalized group n = 20	p
Charlson score, m ± sd [CI 95%]	7.3 ± 1.21 [6.9-7.7]	7.1 ± 1.86 [6.8-7.4]	ns
Number of drugs at entry, m ± sd [CI 95%]	7.9 ± 3.53 [7.7-8.1]	6.8 ± 2.57 [6.5-7.1]	ns
Polymedication (>4), n (%)	23 (82)	16 (80)	
ADL Score (/6), m ± sd [CI 95%]	5.4 ± 0.72 [4.9-5.9]	5.2 ± 0.95 [4.8-5.6]	ns
IADL Score (/4), m ± ds [CI 95%]	3.0 ± 0.87 [2.5-3.5]	2.2 ± 1.28 [1.8-2.6]	**0.025**
Technical aids, n (%)			
Cane only	12 (43)	8 (40)	
Single walker	4 (14)	3 (15)	

Walker and cane	1 (4)	1 (5)	
None	11 (39)	8 (40)	
Fall in the last 6 months, m ± sd [CI 95%]	3.4 ± 3.13 [3.1–3.5]	1.70 ± 0.92 [1.2–2.2]	ns
Polyfalls (>4 falls), n (%)	8 (29)	0 (0)	
Fear of falling (/4) at admission, m ± sd [CI 95%]	2.4 ± 0.91 [1.9–3.9]	1.4 ± 1.14 [1–1.8]	**0.0004**
Quality-of-life scale for the elderly person (/15), m ± sd [CI 95%]	7.6 ± 3.4 [7.4–7.8]	8.0 ± 2.75 [7.7–8.3]	ns
Women, n = 36	7.2 ± 3.27 [6.5–8.0]	8.08 ± 2.99 [7.1–9.0]	
Men, n = 12	9.2 ± 3.9 [7.5–10.9]	7.86 ± 2.48 [6.7–9]	
Weight, m ± sd [CI 95%]	65.0 ± 17.1 [64.9–65.1]	63.0 ± 13.4 [62.9–63.2]	ns
Albumin at admission (g/L), m ± sd [CI 95%]	40.4 ± 3.78 [40.2–40.6]	33.6 ± 5.95 [33.4–33.7]	**0.00006**

Table 8.3. *Clinical and paraclinical characteristics to inclusion*

8.3.2. *Evaluation of inclusion criteria at 3 months and 6 months*

All results for the two groups are presented in Tables 8.4 and 8.5.

	ETP group n = 28	Hospitalized group n = 20	p
0 months			
Fall in the last 6 months, m ± sd	3.4 ± 3.13	1.70 ± 0.92	ns
[CI 95%]	[3.1–3.5]	[1.2–2.2]	
Polyfalls (>4 falls), n (%)	8 (29)	0 (0)	
Fear of falling (/4) at admission, m ± sd [CI 95%]	2.4 ± 0.91	1.4 ± 1.14	**0.0004**
	[1.9–3.9]	[1–1.8]	
Quality-of-life scale for the elderly person (/15), m ± sd	7.6 ± 3.4	8.0 ± 2.75	ns
[CI 95%]	[7.4–7.8]	[7.7–8.3]	
Women, n = 36	7.2 ± 3.27 [6.5–8.0]	8.08 ± 2.99 [7.1–9.0]	
Men, n = 12	9.2 ± 3.9 [7.5–10.9]	7.86 ± 2.48 [6.7–9]	
	ETP group n = 25	Hospitalized group n = 13	p
3 months			
Number of falls, m ± sd	0.4 ± 0.70	0.6 ± 1.65	ns
[CI 95%]	[–0.1–0.9]	[0.3–0.9]	

Polyfalls (>4 falls), n (%)	0 (0)	0 (0)	
Fear of falling (/4), m ± sd	1.5 ± 1.11	1.9 ± 1.20	ns
[CI 95%]	[1.1–1.9]	[1.4–2.2]	
Quality-of-life scale for the elderly person (/15), m ± sd [CI 95%]	10.0 ± 2.44 [9.7–10.3]	8.1 ± 2.29 [7.3–8.9]	0.025
Women	9.9 ± 2.26 [9.2–10.6]	8.6 ± 1.81 [7.7–9.5]	
Men	11.2 ± 2.59 [9.8–12.6]	7.8 ± 3.03 [6.3–9.3]	
6 months			
	ETP group n = 25	**Hospitalized group n = 12**	**p**
Number of falls, m ± sd	0.08 ± 0.28	0.08 ± 0.32	ns
[CI 95%]	[–0.7–0.9]	[–0.7–0.9]	
Polyfalls (>4 falls), n (%)	0 (0)	0 (0)	
Fear of falling (/4), m ± sd	1.0 ± 1.12	2 ± 1.6	0.047
[CI 95%]	[0.6–1.4]	[1.7–2.3]	
Quality-of-life scale for the elderly person (/15), m ± sd [CI 95%]	11.1 ± 2.23 [10.9–11.5]	8.3 ± 2.38 [8.0–8.6]	0.048
Women	10.9 ± 2.01 [10.3–11.5]	8.4 ± 2.07 [7.3–9.5]	
Men	11.8 ± 2.39 [10.4–13.2]	8.0 ± 3.0 [6.5–9.5]	

Table 8.4. *Evaluation of primary and secondary judgment criteria at 0, 3 and 6 months*

		Average at T0	Average at T3	Average at T6	p T0-T3	p T3-T6	p T0-T6
ETP group	**Number of falls**	3.4 ± 3.13	0.4 ± 0.70	0.08 ± 0.28	< 0.05	ns	< 0.05
	Fear of falling	2.4 ± 0.91	1.5 ± 1.11	1.0 ± 1.12	0.001	0.01	< 0.05
	Quality-of-life scale for the elderly person (/15)	7.6 ± 3.4	10.0 ± 2.44	11.1 ± 2.23	0.004	ns	0.0001
Hospitalized group	**Number of falls**	1.70 ± 0.92	0.6 ± 1.65	0.08 ± 0.32	0.0001	ns	< 0.05
	Fear of falling (/4)	1.4 ± 1.14	1.9 ± 1.20	2 ± 1.6	ns	ns	ns
	Quality-of-life scale for the elderly person (/15)	8.0 ± 2.75	8.1 ± 2.29	8.3 ± 2.38	ns	ns	ns

Table 8.5. *Evolution of judgment criteria between inclusion, 3 and 6 months of the two groups (m±sd)*

8.3.3. *Therapeutic reorientation and lost to follow-up*

Between inclusion and the third month, three patients in the ETP group were discharged from the study (one voluntary discontinuation of ETP management and two hospitalizations not caused by a fall). In the Hospitalized group, four patients entered in EHPADs (20%) and three patients could not be reached by telephone due to incorrect administrative information (telephone number). Between the third and sixth months, an entry into an EHPAD occurred in the Hospitalized group. There was no loss to follow-up in the ETP group during this period. The characteristics of lost to follow-up and therapeutic reorientations are described in Table 8.6.

	ETP group n = 3	Hospitalized group n = 8
Cannot be reached by telephone	0 (0)	3 (38)
Voluntary cessation in ETPs	1 (33)	0 (0)
Hospitalization	2 (67)	0 (0)
Entry in EHPAD	0 (0)	5 (63)
Deaths	0 (0)	1 (13)

* Percentage of those lost to follow-up

Table 8.6. *Characteristics of lost to follow-up and therapeutic reorientations*

8.3.4. *Cost evaluation for each branch of the study*

Income and expenditure are presented in the table below. These tables only calculate the difference between expenditure/revenue for 20 patients, but does not assume the profitability of a hospital service.

	Income in euros	Expenses in euros	Differential between income and expenditure
Full hospitalization 20 stays	305,280	119,700	185,580
Day hospitalization 20 admissions	127,440	45,480	81,960

Table 8.7. *Income and expenditure*

8.4. Discussion

This study showed that the group of patients who benefited from therapeutic education had a significantly improved quality of life at 3 and 6 months, associated with a decrease in fear of falling at 6 months and a higher level of home maintenance than the control group. After six months, the number of ETP falls was divided by three despite the predominance of polyfall patient entry. These patients still at home were polypathological and were no less at risk of falling than institutionalized patients. In the literature, studies showing an improvement in the number of falls after educational management exist for patients living in the community (Hill *et al.* 2017; Hill *et al.* 2014; Yoo *et al.* 2013; Hill *et al.* 2015; Pohl *et al.* 2015; Faes *et al.* 2010). Logan *et al.* (2010) in their randomized clinical trial had shown a decrease in the number of falls and hospitalization following management in therapeutic education for a duration of 6 weeks, for patients at high risk of falling who had called an emergency department without being hospitalized following a fall. A home study conducted in the USA came to similar conclusions with a decrease in falls at 6 months after the implementation of multidisciplinary therapeutic education workshops at the patient's home (Bamgbade *et al.* 2016). This decrease in the number of falls and the improvement in the perceived quality of life of patients followed at 6 months suggest that it is important to manage the fall over a prolonged period to obtain better efficacy.

Therapeutic education based on the patient's needs and internal resources (Simon *et al.* 2013; Legrain *et al.* 2014) allows the management of patients with chronic diseases. The specific educational approach has already proved its effectiveness in the follow-up of pathologies such as diabetes and heart failure (D'Ivernois *et al.* 2013; Simon *et al.* 2013; Roussel 2015; Debaty *et al.* 2008). This is an innovative approach to falls, especially since repeated falls are a real chronic disease responsible for many bio-psycho-social complications (Gillepsie *et al.* 2012; Cornillon *et al.* 2002; Puisieux *et al.* 2014).

Through the various workshops, therapeutic education offers a unique forum for individual and group exchange focused on the physical and psychosocial trauma of falling (Juilliard *et al.* 2016). One of the effects of this collective management in therapeutic education may explain the halving of patients' fear of falling in 6 months, while patients who are not being managed see their fear of falling worsen. Our experience shows that the setting up of photolanguage workshops and describing the risk of falling in their domicile allows the patients to share their fear of falling in their daily life (Landrot *et al*. 2007). They discuss their fears spontaneously and freely about the trauma that a fall can cause at home (limitation of activities, changed family outlook). The group's dynamic around this issue enables them to find solutions to adapt their daily lives to the risk of falls by starting from their life history and their needs to become active players in their health (Faes *et al*. 2010; Khong *et al*. 2015). The recovery of confidence in one's physical and psychological capacities through various workshops can restore an image of oneself (Puisieux *et al*. 2014; Lagardère *et al*. 2013). This contributes to statistically improving the quality-of-life scale with consequences in their daily lives (outdoor outings, resumption of safe walking, social exchanges) (Landrot *et al*. 2007).

Six months after a fall, patients not managed in therapeutic education present a worsening of their fear of falling (multiplied by 1.4), a higher rate of entry into institutions (five patients entered EHPAD) with a poor quality of life perceived at QPAM. One of the ways to improve these results would be to take charge, following their return home, of patients in therapeutic education in town medicine, to make the attending physicians aware of the consequences of falls in order to be more systematic in their screening and to integrate therapeutic education in consultation with these elderly patients through short motivational interview techniques while considering the constraints of general practitioners. Involve primary caregivers more where possible. Training in therapeutic education for medical students would be a solution to generalize its use during all consultations (Rey *et al*. 2016).

All these results should be moderated in view of the limited statistical power of the small number of staff. Measures have been taken to avoid bias; however, we cannot exclude possible selection bias represented by populations from different hospitalizations. The number of falls was reported by patients, which can also contribute to a measurement bias that seems difficult to control despite the fact the fall report logs were not always filled out correctly.

However, these results remain very encouraging. They justify the development of therapeutic education for elderly patients at high risk of falls in order to limit the consequences as well as, probably, future hospitalizations and related health costs. They also allow one to reflect on the relevance of new workshops linking general medicine, caregivers and hospital structures.

8.5. Conclusion

The therapeutic education of the patient over 75 years of age living at home is one of the complementary avenues to consider limiting the recurrence of falls and their medico-psychosocial consequences. The follow-up care day hospital is a facilitating place that has made it possible to observe in a population living in a community (average age of 85) the decrease in the number of falls at 6 months, the improvement in the fear of falling and the quality of life of the people followed. The ETP must be integrated into each patient's care pathway to adapt to each patient's real needs, so that together patients and caregivers find solutions to improve medical follow-up, autonomy and quality of life as recommended by the HAS. Faced with the increase in life expectancy, it is necessary to find prevention and care strategies to act as well as possible on the health of elderly people who wish above all to remain at home with a preserved quality of life.

Management in therapeutic education in day hospitals therefore seems to fulfill its role in the prevention of falls. Its access must be extended to as many people as possible. Training and awareness-raising for all doctors and paramedics with city-hospital links should be promoted.

8.6. References

Arveux, I., Faivre, G., Lenfant, L., Manckoundia, P., Mourey, F., Camus, A., *et al.*, (2002). Le sujet âgé fragile. *La Revue de Gériatrie*, 27(7), 569–581.

Bamgbade, S., Dearmon, V. (2016). Fall prevention for older adults receiving home healthcare. *Home Healthcare Now*, 34(2), 68–75.

Beauchet, O., Dubost, V., Revel Delhom, C., Berrut, G., and Belmin, J. (2011). How to manage recurrent falls in clinical practice: guidelines of the French Society of Geriatrics and Gerontology. *The Journal of Nutrition, Health and Aging*, 15(1), 79–84.

Cornillon, E., Blanchon, M.A., Ramboatsisetraina, P., Braize, C., Beauchet, O., and Dubost, V. *et al.* (2002). Impact d'un programme de prévention multidisciplinaire de la chute chez le sujet âgé autonome vivant à domicile, avec analyse avant–après des performances physiques. *Ann Réadapt Médecine Phys.*, 45(9), 493–504.

D'Ivernois, J.-F. and Gagnayre, R. (2011a). *Apprendre à éduquer le patient: approche pédagogique*, 4th edition. Maloine, Paris.

D'Ivernois, J.-F., and Gagnayre, R. (2011b). Compétences d'adaptation à la maladie du patient : une proposition. *Educ Thérapeutique Patient, IPCEM*, 3(2), S201-5.

D'Ivernois, J.-F. and Gagnayre, R. (2013). Éducation thérapeutique chez les patients pluripathologiques: propositions pour la conception de nouveaux programmes d'ETP. *Educ Thérapeutique Patient*, 5(1), 201-4.

Dali, M., Vlasie, M., Daumet-Gonin, A-L., Paul, G., Sauvanaud, F., and Rolland, A. *et al.* (2016). L'ETP influence-t-elle la qualité de vie et les représentations du patient? *Congrès Santé Education Association Française pour le Développement de l'Eduction Thérapeutique.* Maison de la Chimie, Paris.

Dargent-Molina, P. and Breart, G. (1995). Epidémiologie des chutes et des traumatismes liés aux chutes chez les personnes âgées. *Revue d'épidémiologie et de Santé Publique*, 43(1), 72–83.

Debaty, I., Halimi, S., Quesada, J.L., Baudrant, M., Allenet, B., and Benhamou, P.Y. (2008). A prospective study of quality of life in 77 type 1 diabetic patients 12 months after a hospital therapeutic educational programme. *Diabetes Metab.*, 34(5), 507–13.

Faes, M.C., Reelick, M.F., Joosten-Weyn Banningh, L.W., Gier M.D., Esselink, R.A., and Olde Rikkert, M.G. (2010). Qualitative study on the impact of falling in frail older persons and family caregivers: foundations for an intervention to prevent falls. *Aging Ment Health*, 14(7), 834–42.

Gillespie, L.D., Robertson, M.C., Gillespie, W.J., Sherrington, C., Gates, S., and Clemson, L.M. (2012). Interventions for preventing falls in older people living in the community. *Cochrane Database Syst Rev.*, 9,CD007146.

Haute autorité de santé (2009). Recommandations de bonnes pratiques professionnelles: évaluation et prise en charge des personnes âgées faisant des chutes répétées. Report, *Haute autorité de santé*.

Hill, A.-M., Etherton-Beer, C., McPhail, S.M., Morris, M.E., Flicker, L., and Shorr, R. *et al.* (2017). Reducing falls after hospital discharge: a protocol for a randomised controlled trial evaluating an individualised multimodal falls education programme for older adults. *BMJ Open*, 7(2), e013931.

Hill, A.-M., McPhail, S.M., Waldron, N., Etherton-Beer, C., Ingram, K., and Flicker, L. *et al.* (2015). Fall rates in hospital rehabilitation units after individualised patient and staff education programmes: a pragmatic, stepped-wedge, cluster-randomised controlled trial. *Lancet Lond Engl.*, 385(9987), 2592–9.

Hill, A.-M., Waldron, N., Etherton-Beer, C., McPhail, S.M., Ingram, K., and Flicker, L. *et al.* (2014). A stepped-wedge cluster randomised controlled trial for evaluating rates of falls among inpatients in aged care rehabilitation units receiving tailored multimedia education in addition to usual care: a trial protocol. *BMJ Open*, 4(1), e004195.

Insitut national de veille sanitaire (2013). Mortalité par chute accidentelle en France métropolitaine. Report, INVS.

Juillard, S., Paillet, C., Boibieux, O., Beard, M., Daux, E., and Grangette, S. *et al.* (2016). L'éducation thérapeutique: d'autres regards sur la maladie rénale chronique. *Congrès Santé Education Association Française pour le Développement de l'Education Thérapeutique*, Maison de la Chimie, Paris.

Kendrick, D., Kumar, A., Carpenter, H., Zijlstra, G.A.R., Skelton, D.A., and Cook, J.R. *et al.* (2014). Exercise for reducing fear of falling in older people living in the community. *Cochrane Database Syst Rev.*, 11, CD009848.

Khong, L., Farringdon, F., Hill, K.D., and Hill, A.-M. (2015). We are all one together: peer educators' views about falls prevention education for community-dwelling older adults: a qualitative study. *BMC Geriatr.*, 15, 28.

Lagardére, P., Pardessus, V., Beghin, V., Sepieter, C., Petit, V., and Puisieux, F. (2013). Introduire une démarche éducative dans la prise en soin du sujet âgé chuteur. *Rev Gériatrie*. 38(1), 47–57.

Lagger, G., Chambouleyron, M., Lasserre-Moutet, A., Golay, A., and Giordan, A. (2008). Éducation thérapeutique 1re partie: origines et modèle. *Médecine*, 4(5), 223–6.

Landrot, M.D.R., Perrot, C., Blanc, P., Beauchet, O., Blanchon, M.A., and Gonthier, R. (2007). La prise en charge de la peur de tomber apporte-t-elle un bénéfice au patient âgé chuteur vivant en milieu communautaire? À propos d'une étude pilote de 15 cas. *Psychol Neuropsychiatr Vieil.*, 5(3), 225–34.

Legifrance (2015). Arrêté du 14 janvier 2015. Available at : https://www.legifrance.gouv.fr/eli/arrete/2015/1/14/AFSP1501146A/jo/texte.

Legrain, S. (2014). Le programme OMAGE dans une logique de parcours. *AFDET*, France.

Logan, P.A., Coupland, C.A.C., Gladman, J.R.F., Sahota, O., Stoner-Hobbs, V., and Robertson, K. *et al.* (2010). Community falls prevention for people who call an emergency ambulance after a fall: randomised controlled trial. *BMJ*, 340, c2102.

Loi (2009). 2009-879 du 21 juillet 2009. Available at: https://www.legifrance.gouv.fr/affichtexte.do?cidTexte=JORFTEXT000020879475&categorielien=id.

Mauduit, L. (2014). *Aide-mémoire. L'éducation thérapeutique du patient*. Dunod, Paris.

Pin S., Spini D., Bodard, J., Arwidson, P. (2015). Facilitators and barriers for older people to take part in fall prevention programs: A review of literature. *Revue d'épidémiologie et de Santé Publique*, 63(2), 105–18.

Pohl, P., Sandlund, M., Ahlgren, C., Bergvall-Kåreborn, B., Lundin-Olsson, L., and Wikman, A.M. (2015). Fall risk awareness and safety precautions taken by older community-dwelling women and men—a qualitative study using focus group discussions. *PLoS ONE*, 10(3).

Puisieux, F., Lagardère, P., Beghin, V., and Pardessus, V. (2014). La place de l'éducation thérapeutique dans la prise en charge du patient âgé chuteur. *Cah Année Gérontologique*, 6(4), 163–8.

Rey, C., Verdier, E., Fontaine, P., and Lelorain, S. (2016). Renforcer l'implication des médecins hospitaliers en éducation thérapeutique: pistes pour la formation continue et l'accompagnement d'équipe. *Educ Thérapeutique Patient*, 8(1), 10105.

Robinet, P., Puisieux, F. (2014). French geriatric day hospitals managing falls. *Santé Publique Vandoeuvre-Lès-Nancy Fr.*, 26(6), 795–801.

Roussel, A. (2015). Evaluation de l'efficacité d'un programme d'éducation thérapeutique chez les insuffisants cardiaques: le programme I-Carea l'hôpital Saint-André de Bordeaux. Thesis, Université Bordeaux 2, Paris.

Sandrin-Berthon, B. (2009). Education thérapeutique du patient: de quoi s'agit-il? *Actual Doss En Santé Publique*, (66), 10–5.

Simon, D., Traynard P.-Y., Bourdillon, F., Gagnayre, R., and Grimaldi A. (2013). Education thérapeutique: prévention et maladies chroniques. *Elsevier Health Sciences*, 400.

Yoo, J.-S., Jeon, M.Y., and Kim C.-G. (2013). Effects of a fall prevention program on falls in frail elders living at home in rural communities. *J Korean Acad Nurs.*, 43(5), 613–25.

Part 5

The Health Network

Introduction to Part 5

The network is defined as "a frame or structure composed of elements or points, often referred to as nodes or vertices, linked together by associations or bonds, ensuring their interconnection or interaction and whose variations obey certain operating rules"[1]. There are neural networks, spy networks, railway networks, friendship networks and computer networks completed in mathematics by Euclidean networks. In short, anything can be network!

For two decades, healthcare has tried to embark on this path and to introduce it as a new concept of care. The doctor is in touch with the nurse, the dietician and the physiotherapist, and they all are in touch with the patient who is in touch with other patients, all orchestrated by a coordinator who ensures that the whole network works well. The final idea is to show that, thanks to this structured organization, disease is more effectively treated. Many large families have thus been created: the "diabetes network" family, the "cancer network" family, the "cardiovascular risk network" family and the "alcohol network" family, all more distinguished than the others and all imposing themselves with emphasis as tomorrow's health decision-makers.

Why not? The idea of information quickly shared, structured according to needs and carried by many can seduce professionals and patients by reviving a group spirit always more inclined to carry skeptics and melancholic than the solitary destitution of a single medical practice. The problem is that association implies some technical and organizational constraints. As in a circus with multiple artists, we need a ringmaster who animates the different acts without which chaos sets in. The "medical

---

1 Ost, F., and van de Kerchove, M. (2002). De la pyramide au réseau ? Pour une théorie dialectique du droit. Publication des Facultés universitaires Saint-Louis, Brussels.

ringmaster" identifies needs, articulate them and create the necessary links so that everyone can come together around the care program. Then, it is necessary to ensure the sharing and circulation of information. Nothing is better than word-of-mouth so that everything is known, or a few well-meaning busybodies and you will have an exact description of your neighbor's activities. However, in this case, the information must be accurate, legible and comprehensible, with a hierarchy according to the intended recipient. In our computer age, it boils down to the interconnection between professional software and smartphones. A whole language in a computer code such as Python and Java has to be developed and that is not so easy to learn.

Unfortunately, all this has a cost, and in the current context of public financial stagnation, health networks in France are struggling to make their voices heard. Strengthened by statistics and their own "networks of influence", they try to prove their usefulness and ensure their sustainability. However, like the major capitalist economic models, they are being asked to optimize their organizations by diversifying their actions, as well as regrouping and relocating their coordination centers. This logic can be understood because public financing is nothing other than our own purses, and carried by a growing narcissistic instinct and associated taxes, it is very easy to adhere to this demand for reform.

However, even if it becomes urgent to reform the public economic system, this must be done in an enlightened manner without disarticulating organizations whose effectiveness requires a long-term commitment. This is the case of health networks whose impact on a population and public health level can only be recognized after years of work and coordination. A difficult task for an intimacy built over time with a disparate terrain with delicate cohesion.

Technologies can help to centralize information that fortunately remains dependent on holders who can, depending on their mood, cut the digital thread that links them to their distant "Big Brother". The trust that will give meaning to this thread remains a matter for humankind and nothing can replace a relationship woven between two entities that understand each other.

These organizational issues inform the following presentations. Whether it is to set up a network of care with patients or a network of doctors to interpret information about the population they treat, the same constraints of reliability and trust must be established even if current technological revolutions make it possible to rethink the model.

The Evolution of the Economic Model of the Health Network in France: Challenges and Prospects

9.1. Introduction

The "network" structure can be described as agile because of its network architecture, its human dimension and the strategic dynamics of trust and cooperation it promotes (Assens 2005).

In a network, members seek to preserve their capacity to create wealth, by uniting through social cohesion, belonging to a region or promoting "team spirit". The "network" structure combines the stability and strength of a large integrated company with the flexibility and suppleness of a specialized SME.[1] This is due to its capacity to:

– combine the resources held by the different members;

– promote transversal cooperation and to share information held by partners (Assens and Perrin 2011);

– establish agreements that generate consensus in collective action (Assens and Bouteiller 2006);

– obtain the support of the majority of its stakeholders so as to promote its self-sufficiency as in an ecosystem (Moore 1996);

Chapter written by Aline LEMEUR.
1 SME – small and medium-sized enterprise.

– establish a social consensus based on mutual trust that promotes cooperation within the network to the detriment of cooperation with the outside (Assens and Accard 2007);

– base exchanges on a "win–win" or "give-and-take" logic of reciprocity (Axelrod 1980) and to reduce opportunism through the logic of trust and respect for tacit rules.

The network can then be assimilated to a "negotiated environment" in the sense of Pfeffer and Salancik (1978), whose legal, regional or capitalist boundaries are not clearly defined (Assens 2005). The network, which can evolve in a hybrid context, both public and private, relies on tacit agreements between participants to circumvent bureaucratic pressure from public authorities, access majority power or exert lobbying pressure (Cremadez 2004).

By working in networks with its multiple stakeholders (Freeman 1984), the organization improves its environmental monitoring and takes into account societal issues, as well as the financial and economic challenges weighing on its strategy.

However, network agility can be compromised by a multitude of factors, including dependence on external resources. Our choice to focus on this contingency factor is explained by the specificity of the health sector in France, to which we turn our attention in this chapter, which has the particularity of drawing its main financial resources from public funds.

Like the social housing sector, the economic model of health networks is non-profit. It responds to social missions and is expressed by the non-remuneration of directors, the non-distribution of results and the non-integration of results into capital (Hoorens 2013). It is based on a financing plan based on the fundamental principles laid down by laws and regulations, relying mainly on public subsidies allocated in the form of annual grants or budgets.

The 1996 Juppé reform constituted the first regulatory recognition of health network financing, founding the principle of "experimental networks" and introducing new means of regulation. The order 96-345 of April 24, 1996 deals with the medicalized control of healthcare expenditure; it was supplemented at the end of the 1990s by the publication of several texts regulating the methods of financing and operating "networks".

In 1999, "FAQS" or *Fond d'Aide à la Qualité des Soins* (the Quality Care Fund) was created. From 2000 onwards, local health networks financed by the government are bound by a single set of specifications and are required to carry out a self-evaluation accompanied by an external evaluation. Subsequently, the *Objectif National d'Assurance Maladie* (ONDAM, National Health Insurance Objective) continued to supervise the financing of health networks. In compliance with articles L6321-1 of the *Code de la santé publique* (CSP, French public health code) and L 132-43 of the CSS, health networks replacing the "Soubie" networks (article no. L 162-31-1 CSS) and hospital networks (article no. L 6121-5 CSP) are financed through a specific budget. This is the *Dotation nationale de développement des réseaux* (DNDR, French for network development endowment). In order to perpetuate the activity of the health network, a regional allocation under the name of the *Dotation Régionale de Développement des Réseaux* (DRDR, French regional endowment for network development) is created.

In 2007, decree no. 2007-973 of May 15, on the intervention fund for quality and coordination of care, reformed the financing of networks by creating the *Fonds d'Intervention de la Qualité et de la Coordination des Soins* (FIQCS, French intervention fund for quality and coordination of care), which merged FAQS and the DNDR. The FIQCS integrates the *Caisse Nationale d'Assurance Maladie des Travailleurs Salariés* (CNAMTS, French national fund for employee health insurance) and is managed by the *Missions Régionales de la Santé* (MRS, French regional health agencies) and the *Unions Régionales des Caisses d'Assurance Maladie* (URCAM, French regional health insurance funds unions). These funds finance the health networks. This gives institutions the power to select networks for funding based on regional criteria and public health priorities. Thus, the economic model of the health network is a source of dependence.

Indeed, the network's dependence on external resources can make it difficult to maintain unity in diversity and puts the architecture of its network and the rules of the relational game in difficulty (Moreau Deffarges 2003). This can compromise decision-making and the division of tasks or even the way in which the collective interest is reconciled with individual issues, leading to an increase in the heterogeneity of relations (Provan and Kenis 2007).

9.2. Theoretical framework: resource dependence

In an environment that suffers from instability, complexity, diversity and hostility (Mintzberg 1986), some participants can control resources essential to the organization and make it dependent on them. Multiple dependencies give these participants power over the organization, which is then forced to respond to multiple and sometimes contradictory demands (Mintzberg 1982).

Regardless of the organizational structure, influence can be translated in different ways. It takes the form of the control of a vital resource, technical know-how and a body of knowledge crucial to the organization or the control via legal prerogatives (Rojot 2005). Influence can also take the form of formal and direct control over outcomes, informal and targeted control through pressure groups and social norms. Beyond the formal power derived from position in the organizational chart that legitimizes influence, there are implicit power roles linked to the production processes or strategic activities of the organization (Mintzberg 1986). Certain tasks, considered essential in the organization, confer influence on the entities that perform them (Pfeffer and Salancik 1978). Beyond the hierarchical position, the power to influence may be held by certain individuals and/or units who possess skills that are difficult to substitute within the structure. It can also be held by an organization's external environment, if it controls resources that are difficult to substitute.

Thus, a high degree of dependency (or interdependence) increases uncertainty and threatens the survival of the organization, which is forced to manage the demands of the participants on whom it depends and who constitute its environment (de Montmorillon 2002).

A countervailing power is then necessary to manage this dependence, which seems to depend on the degree of non-substitutability of the resource for the organization, the degree of concentration of control over the resource or the degree of discretion over its use and allocation (Mintzberg 1986). Thus, the role of decision-makers would be to work to reduce the causes of dependence of their organization or to increase its degree of dependence on other environmental organizations (Rouleau 2011).

9.2.1. *Dependence-generating factors*

The following four factors can be an essential source of dependence for an organization: the "**scarce**", "**essential**", "**concentrated**" and "**non-substitutable**" nature of the resource, know-how or knowledge (Mintzberg 1986). To be a dependency factor, knowledge, know-how or resources must be essential for the functioning of the organization. They must be rare, non-substitutable, non-replaceable and concentrated in the hands of one participant or a small group of participants who operate together. These include financial resources, expertise, information, contacts, access to senior leaders and so on, where scarcity generates power (Pfeffer and Salancik, 1978). Consequently, the organization becomes dependent on them (Pfeffer 1981).

Beyond the dependence on hierarchical power linked to the position held within the organization, the nature of the tasks performed, where certain tasks are considered essential, is also a source of dependence. The existence of contingencies, linked to the exercise of power, also explains the dependence of certain organizational and operational subunits on others. The degree of interdependence is variable, as units that are difficult to substitute are less dependent and control uncertainty better than others. Power is then contingent on the position occupied by the various units of the organization (Hickson *et al.* 1971).

9.2.2. *Reducing factors of dependence*

Organizations try to overcome their dependency or secure their access to vital resources by trying to influence the constraints and social context in which they operate. They also seek to influence the process of interdependence between organizations, which varies according to available resources in relation to demand. They maneuver to generate uncertainty in a chain and combine their efforts with those of other participants sharing their objectives (Rojot 2005). This gives depth to coalitions, to the political dimension and to the issues of power in the life of organizations (Cyert and March 1963).

The strategy is to find ways to avoid or reduce dependence. The organization may give in to environmental demands, but this may reduce its freedom of action and prove unfavorable to its interests in the long term. It can also avoid environmental influences by responding sequentially to different conflicting demands to partially satisfy them, while playing different interest groups off against each other (Mintzberg 1986). It can affect the conditions of social control in different ways, such as the mobilization of appropriate political games, or even the strategies of individual and collective participants (Crozier and Friedberg 1977). Cross-representations of administrators at the board level also appear to be a useful way to manage resource dependence (Lomi et al. 1999).

In addition, the organization can adapt its structures through vertical integrations, horizontal integrations or mergers and diversification. It can also work to build a "negotiated environment" (Pfeffer and Salancik 1978), negotiating with the environment, coordinating behavior and developing collective structures for interorganizational action to manage mutual interdependencies, through tacit or explicit agreements based on cooperation and reciprocity, "gentlemen's agreements" based on trust and codes of good conduct (Rojot 2005). The network organization is then justified.

The relationship within the network can thus promote control over dependence on resources, thanks to the exchange of information; the establishment of behavioral conventions between members; obtaining support from the environment (suppliers, customers, competitors, etc.), thanks to the culture of partnership; strengthening the effects of experience and learning; mobilizing trust at the expense of negotiation and reducing transaction costs (Assens 2005).

9.3. Empirical study: gerontological health networks in Île-de-France

We complete our theoretical analysis with an empirical survey. We choose as our research field the gerontological health networks in Île-de-France, the region surrounding Paris.

Gerontological health networks have started from the initiative of participants in the field, with the objective of preventing organizational dependency and keeping the elderly and frail at home as long as possible, while controlling costs.

The objective of the gerontological health network is to facilitate access to care and home care for the elderly and frail, thanks to the coordination of experts around them. The choice of network organization in this field is justified by the diversity of the pathologies of the elderly, by the insufficient capacities of traditional structures for the care of frail and dependent older people and by the multiplicity and dispersal of care providers and funding organizations. A flexible structure that encourages cooperation around an elderly and fragile person is then useful.

Currently, gerontological and overall health action is under way. Mutualization is often mentioned as a life-saving solution in several areas, including health. This orientation of public policies foreshadows the mutualization of monodisciplinary health networks into a single multi-thematic network, whose regional coverage is more extensive, thus calling into question the specificity of the themes and territories and risking the loss in terms of proximity to the field and the quality of the service provided to the patient.

Under the impact of this project, the environment of gerontological health networks seems to be becoming more unstable because of the almost injunctive introduction of a large-scale change with "on demand" management (Bartoli 2009) and because of the difficult economic situation. It seems to become more complex because of the need to master an extensive and difficult knowledge, around a multitude of new themes. It seems to become more diverse because of the diversity of patients' needs in complex health situations and the variety of geographical regions covered. These networks seem to be losing favor because of the shortage of public financial resources and difficult relations with the supervisory authority.

9.3.1. *Methodology*

Our knowledge construction is based on a hybrid exploratory survey conducted according to an abductive approach that admits continuous back and forth between theory and the field.

We collected primary data through a qualitative survey of 34 gerontological health networks in Île-de-France and through non-participating observations. We conducted semi-directive interviews based on an interview guide with members of gerontological networks occupying different functions (president, director, doctor, nurse, social worker, occupational therapist, secretary, etc.) and we also met with administrators of the *Fédération des Réseaux de Santé Gérontologiques* (French federation of gerontological health networks) in Île-de-France itself. We then analyzed the data collected through vertical and horizontal analysis grids and interpreted them in the light of the theoretical reading grid around resource dependence. These data that were collected and processed constituted our empirical material and were consolidated and returned.

9.3.2. *Interpretation of results*

Our empirical survey has shown that the power of the external environment is indeed decisive for the life of a health network: their survival depends on it, because of their needs in resources, especially financial. But we have seen maneuvers on their part to protect themselves by developing a counteractive strategy.

9.3.3. *Dependence on the resources of the gerontological health network in Île-de-France*

The external environment of the gerontological health network seems to be dominated by the *Agence Régionale de Santé* (ARS). It exercises formal control over the networks because of the regulatory requirements of the French health system, which links health structures operating in a territory to the ARS in that territory. Beyond formal control, it exercises direct control over their results and even over the use of the resources allocated to them. It thus exercises control over a vital resource for networks: financial resources. The economic model of the health network is entirely based on public money, allocated in the form of a grant or endowment. There is thus a concordance between the real power held and exercised by the ARS and its perceived power (Pfeffer 1981) by the gerontological health networks.

Dependency factors for the gerontological health network in Île-de-France

The power of the ARS vis-à-vis health networks derives in particular from the "rare", "essential", "concentrated" and "non-substitutable" nature of the financial resources necessary for their functioning (Pfeffer and Salancik 1978).

Financial resources are increasingly scarce as a result of economic crisis, which is leading to a scarcity of public resources. This is in addition to the extension of life expectancy, the inversion of the age pyramid with fewer taxpayers and more unemployed, and medical advances that prescribe a multitude of increasingly long and costly medical protocols, for the treatment of various diseases that are now treatable.

Moreover, the concentrated and non-substitutable nature of the financial resources of the health network seems to stem from the specificity of the French health system, which concentrates the financial resources of health structures in the hands of public authorities. The regulations link them to a particular entity of the public health scheme, the ARS, which decides on the allocation of their budget and subsequently controls its use.

However, this dependence is not necessarily inevitable. Our empirical investigation has allowed us to observe different and diversified maneuvers for health networks to protect themselves by developing a counter-strategy.

Factors reducing dependency for the gerontological health network in Île-de-France

With regard to the non-substitutability of public resources, we noted an inequality in the positions of the health networks encountered. Despite the almost absolute power of the ARS and the predominance of public funding for networks, we have seen attempts by some networks to protect themselves and to develop a counter-strategy. To counteract the ARS's injunction to no longer fund certain physician positions, some networks called on hospitals in their sector to make specialized physician time available to them; others called on physicians on their Board of Directors, or on their medical director, who dons two caps (administrative and medical). Some have extended their turnover, externally to all the professional participants in the field, the families of patients cared for by the network and the network's employees internally, because they are seen as sources of wealth and strength, in the face of the supervisory authority. Others have associated themselves with medico-social structures working in the same territory and financed by other public entities, such as their *Conseil régional* (regional council). They then build together a health coordination platform, sometimes even multi-themed, in the image of an empire in the sense of Mintzberg (1986). Some have responded to calls for a European project to free themselves from the influence of the ARS and the ramifications of the French health system, and also some have chosen the path of persistent demands to the ARS, particularly for a territory reduced to a relative health desert.

To cope with the ARS's desire to reduce funding for premises, some networks have also called on the town hall to provide them with premises, while others have shared premises with other health, social or medico-social structures.

To guard against the will of the ARS, which conditioned the financing of the network to the change of its territory of geographical coverage and its themes, certain refractory networks mobilized all the elected representatives of their territories to plead their causes with the ARS. They mobilized against the ARS an "external means of influence", in the sense of Mintzberg (1986), called "lobby campaigns", and then formed themselves as an external coalition around the ARS and sought to influence it.

To avoid the overarching power of the ARS, some networks have anticipated its project, voluntarily opening up to a multitude of themes and/or extending their geographical coverage, even before the multi-thematic networks project was formalized and officially launched. Their strategic intelligence was a facilitator and source of power vis-à-vis the ARS.

The reaction of the ARS then proceeded at several speeds, depending on the network, its territory, its theme(s) or the power of the presidents, Board of Directors, network directors and their regional stakeholders.

Within the gerontological health network, which is tending to become multi-thematic, we are witnessing an exacerbation of political games due to interactions between different doctors and health personnel who are experts in different fields. These include alliance games, games of skill domination and rival camp games in the sense of Mintzberg (1986). The political ability of the network's internal participants seems to have become decisive, in terms of both the internal and external environments. In terms of the external environment, the latter will deploy a multitude of political games to oppose the ARS's legitimate system of influence and its project. Our investigation revealed that certain networks acted like up-and-coming youngsters in the sense of Mintzberg (1986) and even chose to take themselves out of the game, in defiance of the ARS's injunction because they deemed it unbearable and unacceptable.

Thus, the hypothesis of Grenier and Guitton-Philippe (2011) stipulating that even if the logic of mutualization in health policies is administered in an injunctive way, it does not necessarily elicit isomorphic responses at the level of the participants undergoing these projects. Strategic movements seem to be taking hold: on the one hand, new and necessary resources may emerge from consolidation. On the other hand, the participants seem to reinterpret public policy and re-appropriate the consolidation approach, taking advantage of this constraint to revise their institutional project and innovate in their missions.

However, the construction of a "negotiated environment" by adopting the "network" form does not seem to be sufficient for health networks to overcome dependency on external public resources in a sustainable way, while preserving their proximity to the field and the quality of their service to the patient. "Relational capital", fostering privileged links within the framework of partnership and alliance, cooperation and trust (Assens 2003), thanks to the loyalty of relations with the outside (Dyer and Singh 1998), does not seem to be sufficient to control dependence on external resources. The economic model of health networks seems to be unsuited to the current health problem, where needs are increasing while public financial means are decreasing (Cremadez 2014). Health networks seem to create value for the users of their services, but do not generate short-term savings (Frattini and Cremadez 2015), thus initiating the beginnings of a new business system (Nabyla 2011).

9.4. Conclusion

This chapter has sought to question the current economic model of the health network in France, with the objective of identifying tools to sustain its action, in a context of institutional change where public funders find in mutualization a salutary solution to the growing shortage of public financial resources. This orientation of public policies recommends the mutualization of monothematic health networks into a single multi-thematic network with a wider geographical region, calling into question the particularism of the themes and regions. However, the loss of proximity to the ground risks making the network lose its meaning and its reason for being. Our observation was made from the perspective of resource dependence theory. We justified this choice of theoretical framework by the particularism of the French health system, which draws its main financial resources from public funds.

We also conducted an empirical study. We have chosen the gerontological health networks in Île-de-France as our research field. We collected data through a qualitative survey of 34 networks and a federation of health networks in Île-de-France, as well as through non-participating observations.

We found that the complexity of health and social realities in France and the scarcity of public financial resources intensify the risk of conflict because of the existence of a multitude of sometimes contradictory objectives, at the level of the ARS and health networks. Moreover, the complexity seems to be exacerbated by the multitude of partners of which a network is composed, without, however, having legitimate means of pressure against them. Also, the multiplication of communities, types of solidarity and forms of trust in networks, sometimes with different territorial coverage and different needs, to which are added the stakes of medical specialties, the particular culture of the medical profession and competitions between specialists, or even specialties, put in difficulty the rules of the relational game and amplify the intensity of the political game.

We have observed that in order to overcome long-term dependence on external resources, clear at the network level, and in order to perpetuate its activities while preserving its proximity to the field and the quality of its patient service, a change in its economic model seems inevitable. Otherwise, the actions of the health network may become subject to a double impasse with serious consequences: the first being at the level of its financing linked to increasingly scarce public funds, and the second being at the level of the meaning of its missions in the health system, which seems to be based on a misinterpretation. The viability of the economic model of the network then seems linked to the perception, by patients and their entourage, professionals and public authorities, of the value it creates, within the framework of a social and solidarity economy.

Although the size and specificity of our field of study does not allow the generalization of the results, this case study illustrates the main issues common to all health networks in France and, as such, the reflection that it allows can be promising.

9.4.1. Avenues for reflection

To overcome long-term dependence on resources, it might be advisable:

a) To evolve health networks from a spontaneous, peer-to-peer form, where the beneficiaries of services adopt a passive attitude from a financial point of view, and more toward a formalized and voluntarist form, where the beneficiaries of their services are involved, as far as possible and according to their resources, in their financing. And the beneficiary is not only a patient, but also a participant of the French health system. These participants may include the attending physician who calls on the services of the network to assist him in the care of a patient suffering from a complex health situation, the hospital which requests the network's help in organizing the return home and also the regional administrations and the urban communities, regional councils that call on the health network to alleviate situations of medical deserts or to improve health and medico-social services at home in their territories or regional health agencies to conduct a strategic watch where the network will constitute an observatory of health and social needs in a territory. Thus, public funding of health networks no longer remains in absolute terms, but becomes linked to specific services that will be regularly evaluated, for a continuous improvement of the services provided to the patient.

b) To preserve the adaptive potential of networks, by means of an accentuated professionalism breaking with spontaneous and associative voluntarism and by means of a well-established evaluation action; which seems necessary. However, we must avoid falling into exaggeration, where an excessive effort to standardize the operation of the network can prove counter-productive because it is incompatible with its heterogeneous nature.

c) To master the political game in all its dimensions and to mobilize in a dynamic and evolutionary approach different typologies of political games, according to the evolution of the internal and external contexts; which also seems necessary. At this level, the resourcefulness of health networks, manifested in different forms, seems to facilitate.

Moreover, this change in economic model must not obscure respect for the "network culture", or even openness, trust, agility, plasticity and adaptability to the context, where its strength lies not in concentration but in spreading. Consequently, any classic concentration dynamics advocating uniformity will conflict with the culture of the network structure and may generate additional coordination costs.

9.5. References

Assens, C. (2003). Le réseau d'entreprise: vers une synthèse des connaissances. *Management International*, 7(4), 49–59.

Assens, C. (2003). *L'organisation des entreprises: vers une structure en réseau*, e-theque, Paris.

Assens, C. and Accard, P. (2007). La construction d'un réseau. L'union européenne, *Gestion et Management Publics*, 5. Available: http://www.airmap.fr/fr/gestion-management-pwoles-2002-2011/.

Assens, C. and Bouteiller, C. (2006). Mesurer la création de valeur réticulaire. In *Logiques de création : enjeux théoriques et management*, Azan, W., Bares, F. and Cornolti, C. (eds). L'Harmattan, Paris.

Assens, C. and Perrin, C. (2011). L'intelligence économique : une stratégie de réseau pour les enterprises. *Revue Internationale d'Intelligence Economique*, 3(2), 137–151.

Axelrod, R. (1980). Effective choice in the prisoner's dilemma. *Journal of Conflict Resolution*, 24(1), 120–136.

Bartoli, A. (2009). *Management dans les organisations publiques*. Dunod, Paris.

Cremadez, M. (2004). *Organisations et stratégie*. Dunod, Paris.

Crozier, M. and Friedberg, E. (1977). *L'acteur et le système*. Le Seuil, Paris.

Cyert, R. and March, J. (1963). *A Behavioral Theory of the Firm*. Prentice Hall, Englewood Cliffs.

Dyer, J.H. and Singh, H. (1998). The relational view: cooperative strategy and sources of interorganizational competitive advantage. *Academy of Management Review*, 23(4), 660–679.

Frattini, M.-O. and Cremadez, M. (2015). Les réseaux de santé. Working document.

Freeman, R.E. (1984). *Strategic Management: A Stakeholder Approach*. Pitman, Boston.

Grenier, C. and Guitton-Philippe, S. (2011). La question des regroupements / mutualisations dans le champ sanitaire et social : l'institutionnalisation d'un mouvement stratégique ? *Management & Avenir*, 7(47), 98–113.

Hickson, D.J., Hinings, C.A., Lee, C.A., Schneck, R.E. and Pennings, J.M. (1971). Astrategic contingencies theory of interorganizational power. *Administrative Science Quarterly*, 16(2), 216–229.

Hoorens, D. (2013). Le modèle économique HLM. Un modèle à suivre. *Revue de l'OFCE*, 128(2), 73–98.

Lomi, A., Corrado, R., and Sandri, S. (1999). La structure sociale du contrôle des entreprises : partage des membres du conseil d'administration et liens sociétaires. In *L'analyse relationnelle des organisations*, Lomi, A. (ed.). L'Harmattan, Paris.

Mintzberg, H. (1982). *Structure et dynamique des organizations*. Editions d'Organisation, Paris.

Mintzberg, H. (1986). *Le pouvoir dans les organizations*. Editions d'Organisation, Paris.

Moore, J.F. (1996). *The Death of Competition – Leadership and Strategy in the Age of Business Ecosystems*. Harper Business, New York.

Moreau Defarges, Ph. (2003), *La Gouvernance*. PUF, Paris.

Pfeffer, J. (1981). *Power in Organization*. Pitman, Marshfield.

Pfeffer, J. and Slancick, G.R. (1978). *The External Control of Organizations: A Resource Dependance Perspective*. Harper & Row, London.

Provan, K.G. and Kenis, P. (2007). Modes of network governance: structure, management and effectiveness. *The Journal of Public Administration Research and Theory*, 18, 229–252.

Rojot, J. (2005). *Théorie des organisations*, ESKA, Paris.

Rouleau, L. (2011). *Théories des organizations*. Presses de l'Université du Québec, Quebec.

Primary Care Electronic Health Data: Good to the Last Byte

10.1. Introduction

The Commonwealth survey compares 11 countries in relationship to a variety of primary care indicators, including quality and access to care, efficiency, equity and healthy lives. Of the 11 countries, France and Canada rank ninth and tenth, respectively. The cost of care is about the same except in the United States, which is higher. Some of the gap may be explained with the use of electronic medical records (EMR) in the higher-performing countries. This is not only whether or not primary care practitioners have an electronic records system but also whether they are used in a meaningful way to make practices more efficient and to provide better care (Hertle 2015).

Family doctors and general practitioners must begin to think of the patients in their practice as a unique population to whom they are providing medical care. This then means that they are not only managing one patient at a time, but also responsible for the quality of care for that whole population, whether it is the level of preventive care delivered to that population (for example, colorectal cancer screening) or chronic disease care (for example, diabetes). In order to accomplish this, practitioners need data about their patients in some aggregate form and electronic decision aids to help with management. This data can only be available if they can get patient data back from their electronic medical system. There have been a number of calls to improve the data available to primary care practices and the use of decision aids to drive improved healthcare (Ferguson 2012; Bodenheimer 2014).

Chapter written by Richard BIRTWHISTLE.

At present, family physicians, researchers and health system analysts in many countries have no access to this data, which is an essential requirement for improvement at the practice and system levels as well as for researchers.

This chapter describes the efforts made in Canada to have an EMR data repository that is useful for health system monitoring, research, disease surveillance and practice quality improvement.

10.2. The Canadian Primary Care Sentinel Surveillance Network (CPCSSN)

CPCSSN was established in 2008 with funding from the Public Health Agency of Canada as Canada's first multi-disease surveillance system based on primary care EMR data. CPCSSN's objectives are to provide a primary care EMR database for feedback for practitioners about their practice population and practice quality improvement, for research and for disease surveillance (Garies 2017).

It is a network of 12 primary care practice-based research networks at 12 Canadian universities. Each network has recruited primary care practitioners to contribute de-identified patient health data from their EMRs to a regional and central database housed securely at the Centre for Advanced Computing at Queen's University. Although the whole chart is not extracted, the data that *is* extracted includes health conditions and diagnoses; physical measurements such as blood pressure, height, weight and BMI, laboratory results, prescribed medications, allergies, vaccines, medical procedures, risk factors (for example, smoking) and referrals. Sociodemographic information, such as year of birth, sex and postal code, is collected. Social and material deprivation indices are calculated. The database currently has data about 1.6 million patients from more than 1,200 primary care practices in seven provinces and one territory in Canada. The data is extracted quarterly from 12 different electronic medical record vendor systems in more than 180 clinics across Canada, cleansed, coded and mapped to a common database structure. Validated case detection algorithms are run against the data set to identify individuals with a number of chronic conditions (asthma, atrial fibrillation, diabetes, hypertension, osteoarthritis, depression, chronic obstructive lung disease, dementia, Parkinson's disease and epilepsy) (Williamson 2014). Other case definitions are in development. CPCSSN has obtained Research Ethics Board (REB) approval from 12 universities and the Health Canada REB to collect this data without direct patient consent.

Legislative privacy requirements are met in all provinces. CPCSSN data managers have developed algorithms to identify and remove potential identifying information in the data. CPCSSN has established procedures to ensure data availability to all researchers.

10.3. EMR data limitations

While EMR data is a rich repository of longitudinal patient health data, it is not without its limitations. Clinical EMR data is collected by the practitioner for the purposes of patient healthcare. As with any data being used for a secondary purpose, they require a critical review and cleansing to make the data fit for purpose. There are three main limitations to the extraction of usable information from EMRs. The first limitation is the variability of the input of the data, such as differences in where patient information is entered into the EMR (for example, smoking history may be entered as social history, a risk factor or as part of the encounter note). Another example is the many ways for practitioners to characterize whether a patient smokes tobacco or not. In order to deal with this, CPCSSN has developed a standard coding to indicate whether the patient currently smokes, was a previous smoker or never smoked. Further, some electronic record systems have better background disease coding than others. There is no expectation by the network that the practitioner codes any data so we must do that in the background. If a practitioner uses short forms such as "DM", "T2DM" or "Diab" for the diagnosis of diabetes mellitus, data managers have developed synonym lists that are all coded to ICD9 code of "250" for diabetes mellitus. The second limitation is data that is difficult to code or analyze, such as scanned documents or encounter notes. The third limitation is that some of the data needed for population surveillance or research is missing or poorly represented (for example, important risk factors and modifiers of chronic disease, such as ethnicity, education and income). Solutions to these problems include providing a more structured data entry that has less free text and choosing an EMR capable of communicating with other electronic systems (for example, in hospitals, laboratories and imaging) to allow direct storage of as much clinical information as possible in the EMR. CPCSSN data managers have spent many hours cleansing data such as this to enhance the quality of the data and it is an ongoing process.

In Canada, the use of CPCSSN data for disease surveillance representativeness of the patients, practices and primary care health professionals is an issue. CPCSSN has assessed this and while not representative of the Canadian general population, the data does represent the population that attends primary care practice (Queenan 2016). The CPCSSN data has older patients and fewer young to middle-aged males. The practitioners are only those who use electronic medical records, are younger and include more women than the population of primary care practitioners. It also includes fewer international medical graduates

10.4. Chronic disease surveillance

CPCSSN data is being used for the surveillance of chronic diseases. Examples of the use of the data are found in a series of publications related to diabetes, COPD, dementia, depression, osteoarthritis and obesity (Bodenheimer 2014; Wong 2014; Godwin 2015; Rigobon 2015; Birtwhistle 2015; Green 2015; Drummond 2016).

Obesity rates in the primary care patient population in CPCSSN were analyzed, as shown in Figure 10.1, and are found to be consistent with other estimates of obesity in Canada. The advantage of using CPCSSN data is that there is longitudinal data in routinely collected secondary data. This is a much less expensive approach than national patient surveys or smaller research studies.

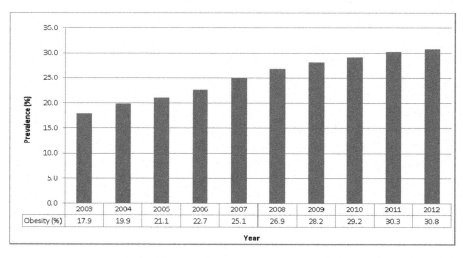

Figure 10.1. *CPCSSN prevalence of obesity in Canada 2003–2012 (Rigobon 2015)*

10.5. Research

CPCSSN data is available for researchers to use. A process requiring review and approval from the CPCSSN Surveillance and Research Committee is in place, which tracks research ethics board approval and how the data is being used. There have been applications for a wide range of data use, including chronic pain, chronic pelvic floor dysfunction, hypertension and chronic renal disease. Increasingly, CPCSSN data is being linked to administrative data such as the hospital discharge database, national ambulatory care records and dispensed drug databases held by provincial health data repositories. An example of this type of research is hospital and emergency room utilization in relation to diabetes control (Birtwhistle 2017).

Furthermore, there is an opportunity to use machine learning, natural language processing and other "big" data analytic approaches to EMR data to look for new associations of chronic disease to novel risk factors and linkage to genomic or proteomic databases that are developing.

10.6. Quality improvement and practice feedback

One of the opportunities for EMR data collection is to be able to provide information back to practitioners about their practice population and chronic disease management in comparison with peers. A major deficiency in EMRs in Canada and elsewhere is that although a practitioner can enter data easily into an EMR, it is almost impossible to get any data back. CPCSSN has developed two approaches to facilitate access to practice data. One is a static quarterly report that is sent to all participating practitioners. It has information about their practice profile and the prevalence of chronic diseases. It also gives more details about how well the blood sugar, blood pressure and cholesterol values in their patients with diabetes are controlled. The report also gives an anonymous comparison of these values to other practitioners in their region and country.

The other tool CPCSSN is using for practice feedback is much more dynamic. It is a password-protected web-based tool called the Data Presentation Tool (DPT) (Greiver 2015), which provides the practitioner or practice analyst with access to their own processed CPCSSN data and a simple query structure to investigate any questions or interests they have in their own practice. This not only provides a useful report, but also allows practitioners to re-identify their patients who they may want to contact to have an appointment for further follow-up. An example of this would be to identify all those people with diabetes with an indication of high hemoglobin A1c (a blood test indication degree of control of diabetes) who have not had a follow-up in the last year. This tool allows for a meaningful use of the practice data, which is unavailable without using considerable administrative time doing manual chart reviews. The tool also allows for custom searches. A recent addition is the deprivation index, which is an area-based postal code proxy for social and material deprivation allowing practitioners to assess the socioeconomic status of their practice population and those with particular chronic diseases. Using GIS technology, we are able to receive a mapping of the geographical distribution of their patients.

Although there is still work to do to make EMRs a valid and reliable source of data in Canada, they can now provide useful data for primary care research and surveillance, make a valuable contribution to the understanding of chronic disease management in primary care, help practitioners provide population management and improve the health of Canadians (Birtwhistle 2015).

10.7. References

Birtwhistle, R., Green, M.E., and Frymire, E. *et al.* (2017). Hospital admission rates and emergency department use in relation to glycated hemoglobin in people with diabetes mellitus: a linkage study using electronic medical record and administrative data in Ontario. *CMAJ Open*, 5(3), E557–E564.

Birtwhistle, R., Morkem, R., and Peat, G. *et al.* (2015). Prevalence and management of osteoarthritis in primary care: an epidemiologic cohort study from the Canadian Primary Care Sentinel Surveillance Network. *CMAJ Open*, 3(3), E270–275.

Birtwhistle, R., Williamson, T. (2015). Primary care electronic medical records: a new data source for research in Canada. *CMAJ*, 187(4), 239–240.

Bodenheimer, T., Ghorob, A., Willard-Grace, R., and Grumbach, K. (2014), The 10 building blocks of high-performing primary care. *Ann Fam Med*, 12(2),166–171.

Drummond, N., Birtwhistle, R., Williamson, T., Khan, S., Garies, S., Molnar, F. (2016). Prevalence and management of dementia in primary care practices with electronic medical records: a report from the Canadian Primary Care Sentinel Surveillance Network. *CMAJ Open*, 4(2), E177–184.

Ferguson, T.B. (2012). The Institute of Medicine committee report "best care at lower cost: the path to continuously learning health care". *Circ Cardiovasc Qual Outcomes*, 5(6), e93–94.

Garies, S., Birtwhistle, R., Drummond, N., Queenan, J., and Williamson, T. (2017). Data resource profile: national electronic medical record data from the Canadian Primary Care Sentinel Surveillance Network (CPCSSN). *Int J Epidemiol.*

Godwin, M., Williamson, T., and Khan, S. *et al.* (2015), Prevalence and management of hypertension in primary care practices with electronic medical records: a report from the Canadian Primary Care Sentinel Surveillance Network. *CMAJ Open*, 3(1), E76–82.

Green, M.E., Natajaran, N., and O'Donnell, D.E. *et al.* (2015). Chronic obstructive pulmonary disease in primary care: an epidemiologic cohort study from the Canadian Primary Care Sentinel Surveillance Network. *CMAJ Open*, 3(1), E15–22.

Greiver, M., Drummond, N., Birtwhistle, R., Queenan, J., Lambert-Lanning, A., and Jackson, D. (2015). Using EMRs to fuel quality improvement. *Can Fam Physician*, 61(1), 92, e68–99.

Greiver, M., Williamson, T., and Barber, D. *et al.* (2004), Prevalence and epidemiology of diabetes in Canadian primary care practices: a report from the Canadian Primary Care Sentinel Surveillance Network. *Can J Diabetes*, 38(3), 179–185.

Hertle, D. and Stock, S. (2015). Commonwealth fund survey 2012: survey of primary care doctors in 11 countries: use of health information technology and important aspects of care. *Gesundheitswesen*, 77(8–9), 542–549.

Queenan, J.A., Williamson, T., Khan, S. *et al.* (2016). Representativeness of patients and providers in the Canadian Primary Care Sentinel Surveillance Network: a cross-sectional study. *CMAJ Open*, 4(1), E28–32.

Rigobon, A.V., Birtwhistle, R., Khan, S. *et al.* (2015). Adult obesity prevalence in primary care users: an exploration using Canadian Primary Care Sentinel Surveillance Network (CPCSSN) data. *Can J Public Health*, 106(5), e283–289.

Williamson, T., Green, M.E., Birtwhistle, R. *et al.* (2014). Validating the 8 CPCSSN case definitions for chronic disease surveillance in a primary care database of electronic health records. *Ann Fam Med*, 12(4), 367–372.

Wong, S.T., Manca, D., Barber, D. *et al.* (2014). The diagnosis of depression and its treatment in Canadian primary care practices: an epidemiological study. *CMAJ Open*, 2(4), E337-342.

Conclusion

This book concludes 4 years of fruitful meetings between engineers and health professionals. From this already well-arranged score, it is to be hoped that other symphonies and concertos will be born to enrich the great field of knowledge and that this multi-cultural orchestra will survive the narcissistic temptation of knowledge. Cooperation can only improve the evolution of a health system in the midst of an existential crisis, and everyone's expertise remains essential to the harmonious functioning of the whole.

Modern Bach continues to dazzle generations of musicians with his multiple voices, Schubert magnifies the instrumental trio, Beethoven the orchestration and Debussy the expressionism, all organizing violins, pianos, cellos and horns to fill the void of our emotional imagination and contribute to its development! The engineer, the computer scientist, the doctor, the nurse and the financier are simultaneously quality instruments and musicians capable of the virtues of composition, and professionals capable of enriching the great repertoire of health, as the experiences presented in this book have shown.

However, even if the great composer, the "Government", remains necessary to create music, health professionals must remain the masters of their playing and impose the tonalities, which make up all the intelligence and the sensitivity of an interpretation.

And where are the patients in all this? They are at the center of attention and should not be offered as fodder to a vibrant science of passion at the risk of forgetting the very meaning of their lives. There is no more beautiful music than that which energizes the weakest, makes the strongest philosophize and raises the multitude!

List of Authors

Virginie ANDRÉ
GICC (CNRS UMR 7292)
Tours University Hospital
France

Vincent AUGUSTO
Team I4S
Center for Biomedical and
Healthcare Engineering
École des Mines de Saint-Étienne
Institut Mines Telecom
France

Mireille BATTON-HUBERT
Team MOE
Institut Henri Fayol
École des Mines de Saint-Étienne
Institut Mines Telecom
France

Jean-Charles BILLAUT
LIFAT (EA 6300)
ROOT (CNRS ERL 6305)
Polytechnic University of Tours
France

Richard BIRTWHISTLE
Center for Studies in
Primary Care
Queen's University
Kingston
Canada

Marie-Ange BLANCHON
Saint-Étienne University
Hospital Center
France

Thomas CELARIER
Saint-Étienne University
Hospital Center
France

Mario DEBELLIS
Réseau de Santé CAP2S
Saint-Étienne University
Hospital Center
France

Justine DIJON
Saint-Étienne University
Hospital Center
France

Maria DI MASCOLO
G-SCOP (CNRS UMR 5272)
Grenoble
France

Mario DI PALMA
Institut Gustave Roussy
Villejuif
France

Thomas FRANCK
Team I4S
Center for Biomedical and
Healthcare Engineering
École des Mines de Saint-Étienne
Institut Mines Telecom
France

Alexandra GENTHON
Plateforme CPS – Coordination
Proximité Santé
Grenoble
France

Régis GONTHIER
Saint-Étienne University
Hospital Center
France

Yannick KERGOSIEN
LIFAT (EA 6300)
ROOT (CNRS ERL 6305)
Polytechnic University of Tours
France

Aline LEMEUR
Higher Institute of Management
(ISM)
Versailles Saint-Quentin-
en-Yvelines University
France

Laure MARTINEZ
Saint-Étienne University
Hospital Center
France

Gérard MICK
GCS Maison des Réseaux de
Santé Isère
Grenoble
France

Robert PICARD
CGEIET
Ministère de l'économie, de
l'industrie et de l'emploi
Paris
France

Michel SABY
Association of Management of
Health Centers of Grenoble
(AGECSA)
and
GCS Maison des Réseaux de
Santé Isère
Grenoble
France

Marianne SARAZIN
Team I4S
LIMOS (CNRS UMR 6158)
Center for Biomedical and
Healthcare Engineering
École des Mines de Saint-Étienne
Institut Mines Telecom
and
Réseau sentinelles
UMR-S 1136 INSERM
Mutualiste Group
of Saint-Étienne
France

Radia SPIGA
Saint-Étienne University
Hospital Center
France

Jean-François TOURNAMILLE
GICC (CNRS UMR 7292)
Tours University Hospital
France

Marc WEISSMANN
Plateforme CPS – Coordination
Proximité Santé
Grenoble
France

Index

Printed in the United States
By Bookmasters